Electric Space

Space-based Solar Power Technologies & Applications

Unlimited Energy from Space

Danny Royce Jones, Ali Baghchehsara

This Publication is in copyright. Subject to statutory exception and to the provisions of relevant collective licensing agreements, no reproduction of any part may take place without the written permission of Authors (Danny Royce Jones, Ali Baghchehsara).

Copyright © 2013 by Danny Royce Jones, Ali Baghchehsara

All rights reserved.

ISBN: 1-4942-5780-7

ISBN-13: 978-1494257804

Library of Congress Control Number: 2013921965

CreateSpace Independent Publishing Platform, North Charleston, SC

First published 2013

Authors has no responsibility for the persistence or accuracy of URLs for external or third-party internet websites referred to in this publication and does not guarantee that any content on such websites is or will remain, accurate or appropriate. Information regarding prices, travel timetables and other factual information given in this work is correct at the time of first printing but Authors does not guarantee the accuracy of such information thereafter.

Printed in United States of America

Dedication:

This Book is dedicated to the people of planet Earth who will soon become a space fairing civilization.

Danny Royce Jones
Ali Baghchehsara

Table of Contents

CHAPTER ONE: WHAT IS SPACE SOLAR ENERGY? 1
CHAPTER TWO: CAN SPACE ENERGY SAVE PLANET EARTH? 6
   SCALE OF GLOBAL ENERGY USE 13
CHAPTER THREE: SPACE-BASED SOLAR POWER SATELLITES 18
   ADVANTAGES OF SPACE SOLAR POWER 21
   DISADVANTAGES OF SPACE SOLAR POWER 22
   WIRELESS POWER TRANSMISSION 23
   MICROWAVE POWER TRANSMISSION 25
   RADIATION AND THE INVERSE SQUARE LAW 27
CHAPTER FOUR: BRIEF HISTORY OF THE SOLAR POWER SATELLITE CONCEPT ... 30
   THE SUN TOWER 32
   SUN TOWER IN LEO 1,000KM 35
   SUN TOWER IN MEO 36
CHAPTER FIVE: TECHNICAL PROBLEMS 37
   SHADOWING PROBLEM 38
   BEAM TIME PROBLEM 38
   OCEAN PROBLEM 39
   SPACE LAUNCH PROBLEM 40
   THE MASS PROBLEM 41
   TRANSMITTER MASS PROBLEM 42
   POWER PRODUCTION PROBLEM 44
   THE GEO PROBLEM 44
CHAPTER SIX: NEW TECHNOLOGY 46
   MASSIVE SOLAR CONCENTRATION 48
   RAINBOW CONCENTRATORS 52
   PHOTOVOLTAIC MASS REDUCTION 57
   PHOTOVOLTAIC COOLING 59
   THE HOLY GRAIL OF SSP 60
   CARBON/CARBON RADIATOR 61
CHAPTER SEVEN: ALTERNATIVE ORBITS 64
   GEO 73
   LOW EARTH ORBIT 76
   MOLNIYA 82
   ELLIPTICAL ORBITS 89
   SPS IN ELLIPTICAL ORBIT 89
   MOLNIYA SATELLITE ASSUMPTIONS 91
   SS-O SUB-MOLNIYA 92

NEW REFERENCE DESIGN 95
MEO THE SWEET SPOT (10,300KM) 96
MEO SS-O 97
POWERSAT SIZE COMPARISON 100
LAUNCH COSTS 101

CHAPTER EIGHT: THE SPACE GRID 103
SUN-SYNCHRONOUS ORBITS 105
SPACE POWER RELAY 109
HOW THE SPACE GRID WORKS 110
SPACE GRID CONSTELLATIONS 119
USING THE SPACE GRID FOR SPACE SETTLEMENT 120

CHAPTER NINE: SOLAR ELECTRIC PROPULSION (SEP) 123
EP SYSTEMS 125
ELECTRIC THRUSTERS 126
SOLAR ELECTRIC SPACE TUG 128
BEAMED ENERGY IN-SPACE TRANSPORTATION SYSTEM 128
GRIDDED ELECTROSTATIC ION THRUSTER RESEARCH 132
HYBRID WPT/SOLAR SPACE TUG 134
SOLAR ELECTRIC SPACE TUG 139

CHAPTER TEN: LARGE SCALE SPACE SETTLEMENT 152
RECTENNA IN SPACE 156
POWERING SPACESHIPS 156
MARS SUPER CHEAP 158
SSO POWERSAT 162
POWERSAT ORBIT DETAILS 163
MICROWAVE GENERATOR 164
ENOUGH ENERGY FOR THE PLANETS? 165

CHAPTER ELEVEN: THE BUSINESS CASE 166
THE MONEY 167
PEAK POWER 168
FINANCING 169
GROUND ASSETS 173
POTENTIAL BENEFITS OF COMMERCIALIZATION 174

CHAPTER TWELVE: CONCLUSION 178

AUTHOR BIOGRAPHY 182

REFERENCES 184

ACCRNYMS 186

ACKNOWLEDGMENTS

A lot of people have worked on the idea of making Space Solar Power a viable technology over the past forty years, including many at NASA and the DOE. Now with new and better technologies interest continues to grow Worldwide and many more are now looking at this subject. There has been substantial interest in Japan and now India is advancing Space Solar Power for benefit of all mankind and every ones efforts have moved us all closer to the day of unlimited clean energy.

Chapter One

What is Space Solar Energy?

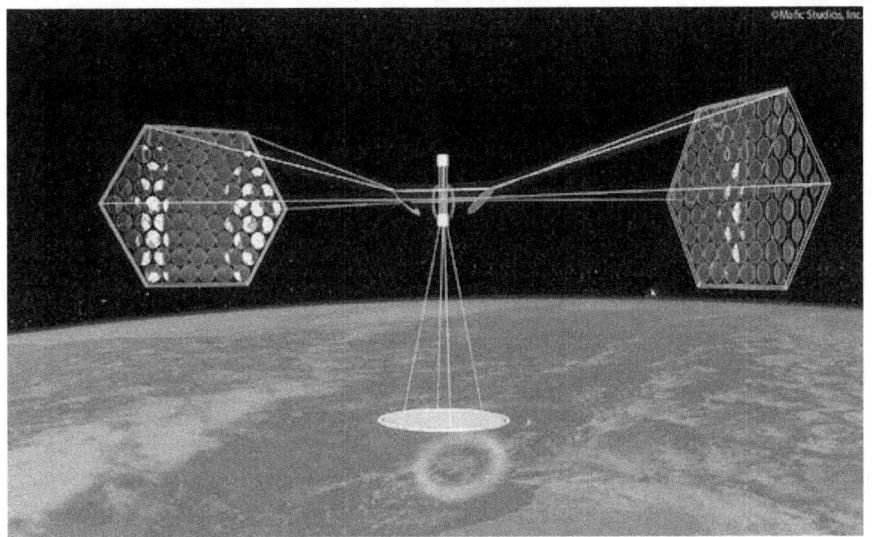

Space Solar Energy using the Integrated Symmetrical Concentrator

Innovation has delivered humanity from the caves to the ability to access the edge of space and to send robotic explorers deep into the solar system. It is the driving force in mankind's expansion, growth and technological development. Beamed energy technology could potentially lower the cost of access to space and transportation though space, thereby lowering the cost of transporting goods and people to many destinations in the solar system and beyond. Technology spin-off from space technology investments will benefit all of humanity. Spin-offs could have a huge positive impact in the areas of electronics and instrumentation, automotive, energy (especially clean energy) and lower cost space transportation. In order for humanity to move into space to settle this new frontier innovation in beamed energy technologies will be required.

We will look at the topic of space energy and the technologies that can be used to generate this energy. While some of the ideas in this book go back to the original concepts many of the ideas and concepts are new. It is hoped that others will apply these ideas into even new and better approaches so that we can move into space in a really big way, but also do this in ways that are cost effective and even profitable. Access to energy in space is the key to developing its resources and those resources are required to support large scale space settlement. We don't need to wait decades to develop the technologies for space energy because these technologies already exist. We are just not using them. These technologies can provide us with the means to travel rapidly and cheaply though space. They can also be used to power our industry and homes on Earth.

Space settlement and space resource management offers us the opportunity to expand our civilization to a new level. Once our civilization moves into space in a way that is sustainable our capabilities will grow in ways never imagined. We will colonize and terraform planets and travel to other star systems. While this may seem like fantasy to some the truth is we already have the basic technology to do these things. We have only to utilize the technology.

The development of space will depend on the availability of energy and lots of it. Once we have mastered that, all things become possible. We will be able to transport people and cargo to many destinations in our solar system, power the earth with clean and abundant energy, and tap the resources of asteroids and comets. Once we have settled the near Earth areas of the solar system, such as, the Moon, Mars and Venus we can strike out to other star systems.

Accelerating the development of space energy is important to the future of mankind's ability to sustain growth and for the protection of the Earth's environment. Space Energy can offer cost effective supplies of power if combined with innovative approaches to in-space power production.

Space Solar Energy is a concept invented by Dr. Peter Glaser in 1968. The idea was to place PowerSats in orbit around the Earth. These PowerSats would collect the Sun's energy and beam the energy down to Earth in the form of microwaves. The problems with the concept includes: the mass of the PowerSats and the ability to launch this mass into orbit. Technological progress since that time has been moving us ever close to the ability to make this dream a reality.

In 1968, Doctor Peter Glazer patented the idea that if you collected solar energy in space using a large satellite and beamed the energy to Earth using microwaves you could provide the Earth with lots of energy. He was issued a patent in 1973. Although Dr. Glaser is credited with inventing the basic concept, it had been popularized by science fiction authors such as Murray Leinster, Olaf Stapledon, Isaac Asimov and Clifford Simak for at least a generation.

In 1941, science fiction writer Isaac Asimov published the science fiction short story "Reason", in which a space station transmits energy collected from the sun to various planets using microwave beams.

This idea became known as Space-based Solar Power or SBSP for short. SBSP, later shortened to just Space Solar Power or SSP, is generally considered to be the collection in space of energy from the Sun and its wireless transmission from space using microwaves or lasers for use on Earth. It has been observed that the implementation of such a system could potentially offer energy security, environmental, and technological advantages if such a system was developed.

It is also interesting to note that this same technology would be very beneficial for in-space transportation and space development. Solar electric propulsion (SEP) has the same basic requirements as SSP; these are large power generation and low mass. The primary difference between SSP and SEP might be the different microwave or laser frequencies used for each application.

This is because in space you don't have the atmospheric losses you have while beaming through the Earth's atmosphere. Wireless power transmission (WTP) allows you to beam energy over very long distances. This is especially useful in space were the distance can be very long.

Glaser's idea could become the next level of alternative energy source that would become a potential replacement of fossil base fuels including coal, petroleum and natural gas.

And as estimated by scientists, by end of 21st century, all the fossil fuels are expected to be fully consumed meaning there will simply not be anymore. This simple fact seems to be largely ignored by most of society. This is not a "global warming" argument, it is a - we will run out of energy argument. Long before that day gets here the stress on energy prices will force a reshaping of the world's wealth between those that have and those that don't have access to carbon rich fuel sources. Current alternative energy sources, such as wind, solar and ethanol can slow down this process, but they can't stop it.

Even combined they cannot produce enough energy to replace what is being depleted. Space Solar Energy is the only technology with the potential to replace carbon fuels at large scale and do so at a cost that is affordable to most everyone on Earth. Unlike carbon fuels like oil, gas, ethanol, and coal plants, space solar energy does not emit greenhouse gases. With Space Solar Energy you will never have to worry about running out of energy, unless of course your PowerSat breaks down.

We have an immense fusion reactor in our sun working for us, this energy source is ultimately responsible for almost all our energy choices, all we have to do is make better use of it by tapping into it more directly.

The concept of generating solar power in space for wireless transmission to receivers on the ground has been discussed at some length during the past four decades. Space Solar Energy represents an attempt to provide humanity with an environmentally benign terrestrial electric power source. In order to satisfy these goals, the solar power generation system must provide very high power levels to the beam power source to maximize delivered power to Earth, be very lightweight to control launch costs, and be extremely long-lived to minimize maintenance and upkeep requirements.

Besides the cost of implementing such a system, SBSP also introduces several new hurdles, primarily the problem of transmitting energy from orbit to Earth's surface for use. Since wires extending from Earth's surface to an orbiting satellite are neither practical nor feasible with current technology, SBSP designs generally include the use of some manner of wireless power transmission. The collecting satellite would convert solar energy into electrical energy on board, powering a microwave transmitter or laser emitter, and focus its beam toward a collector (rectenna) on the Earth's surface. Radiation and micrometeoroid damage could also become concerns for SBSP.

The single biggest problem for space solar power has been and continues to be the mass of the PowerSats. In this book we will explore a number of ways to reduce the size and mass of PowerSats in order to reduce costs and accelerate development of this concept.

Chapter Two

Can Space Energy Save Planet Earth?

As the World's need for energy continues to grow driven by population growth and increased industrialization, new options for energy will be needed. This is especially true given that reserves of conventional energy sources such as oil, gas and uranium are rapidly being depleted and that greenhouse gases being released into the atmosphere are causing large scale damage to the Earth. We find that alternatives energy sources such as wind, ground solar and biomass have physical site limitations and while they are certainly desirable from an ecological prospective, are often high cost and cannot meet the Earth's future energy needs even if fully developed.

---

[1] Image credit: sistergirlnews.com

Is there a source of green energy that is abundant and cost effective? Actually, yes, there is such a resource and it is called Space Solar Energy. Science has long predicted that global warming would cause melting of the north and south poles. It is now common knowledge among people that this is taking place. Islands are disappearing, coast lines are moving inland reclaiming millions of acres back to the seas. They also predicted that it would cause the release of massive amounts of methane gas locked up beneath the ocean floors and that the release of this methane gas would permanently alter our planet with a runaway greenhouse gas effect that would turn Planet Earth into another Venus.

Recently, huge plumes of methane gas were found venting under Siberia. Such a planet could not house life – any form of live. Simply put, the planet Earth will die. Now we don't want to be all doom and gloom so we will focus on the positive by saying that we do have a viable solution available if we choose to take advantage of it.

Space Solar Energy has the potential to cost effectively meet all of the Earth's current and future energy needs. Space Solar Energy also has the potential to reverse America's half a trillion dollar a year balance of payments deficit and to generate a new generation of American jobs. Space Solar Energy is a source whose basic technology is already here.

The United States has been harvesting solar power in space and transmitting it to earth since 1962, when Telstar, the first commercial satellite, launched into orbit. Telstar, looked like a beach ball encrusted with square medallions. The medallions were photovoltaic panels. Today harvesting energy in space and transmitting it to earth is a quarter of a trillion dollar industry…the commercial satellite business. You use solar energy harvested in space when you tune into satellite TV or satellite radio, when you use the Global Positioning System (GPS), when you consult the pictures in Google Earth and when you use your cell phone. The technology now exists to solve the Earth's energy problems with low cost power from space and at the same time solve the Global Warming problem.

This technology will stimulate new investment in launch vehicles and can provide a basis for development of solar electric space transportation. There is still much work to do but there is no longer any question that Space Solar Energy can be an economically viable part of the Earth's energy equation and will be a major factor in moving humanity into space. Space Solar Energy is the next step in space development and it is a step we can take now. I know some people will be skeptical and that is certainly justified since up to now the skeptics have been right, but our technology has now surpassed the skepticism and we will show this as you read on.

The reservoir of Space-Based Solar Power is almost unimaginably vast, with room for growth far past the foreseeable needs of the entire human civilization for the next century and beyond. In the vicinity of Earth, each and every hour there are 1.366 GWs of solar energy continuously pouring through every square kilometer of space. If one were to stretch that around the circumference of geostationary orbit, that 1 km-wide ring receives over 210 terawatt-years of power annually.

The amount of energy coursing through that one thin band of space in just one year is roughly equivalent to the energy contained in ALL known recoverable oil reserves on Earth (approximately 250 terawatt years), and far exceeds the projected 30TW of annual demand in mid-century. The energy output of the fusion-powered Sun is billions of times beyond that, and it will last for billions of years—orders of magnitude beyond all other known sources combined.[2]

---

[2] According to an October, 2007 US Department of Defense (DOD) National Space Security Office (NSSO) study which included representatives from DOE/NREL, DARPA, Boeing and Lockheed-Martin: "A single kilometer wide band of geosynchronous earth orbit experiences enough solar flux in one year (approximately 212 terawatt-years) to nearly equal the amount of energy contained within all known recoverable conventional oil reserves on Earth today (approximately 250 TW-yrs)."

In 2010 Bill Gates suggested that technology can be used to solve the Earth's energy and environmental problems, and specifically climate change, although he didn't mention Space Solar Energy as an option. The authors go further, and suggest that Space Solar Energy technology can not only reduce or eliminate climate change it can potentially supply our planet and the solar system with unlimited quantities of green, cheap energy.

The availability and use of renewable energy sources compatible with reducing risks to the global environment is the key to sustainable development both on Earth and in space. Space Solar Energy is an interesting subject in many ways as the requirements to achieve it are fairly substantial. In order for Space Solar Energy to provide clean, zero emission electrical power to the Earth the first steps are to demonstrate that it is actually technically and economically viable and can be deployed in a cost effective manner.

Space Solar Energy has been considered the energy technology of the future. But what if that future had finally arrived? What if Space Solar Energy was both technically and economically viable now? That is the purpose of the book, to show you that the basic technology for affordable Space Solar Energy is available today.

> "The commercial development of the near space environment has been a dream of many people, companies, and organizations for over fifty years. The limiting factor has been and continues to be the high cost of launching people and payloads into orbit. It can be safely argued that development of near space, or beyond will not happen without large reductions in the cost of launch services. The dream of commercial space development may finally be coming to reality, spurred by incremental advances in technology and the head strong push by small groups of space enthusiast. New-Space space companies are helping to cut the cost of space transportation and move us toward the day when mankind can move into the heavens." [3]

Launch costs are not the only problem though. What is missing is the

---

[3] 1. Gates, Bill; TED 2000, FILMED FEB 2010 ,TED2010, http://www.ted.com/talks/bill_gates.html

right space project that can stimulate new launch systems development. The project should be one that has huge revenue potential and this is where Space Solar Energy fits in. Space Solar Energy is the only near term, large scale space technology that actually has the potential to pay for itself. In 2008, total worldwide energy consumption was 474 Exajoules (474×1018 J=132,000 TWh). There is no clean energy source on planet Earth that can deliver that level of base-load power.

The Sun is a giant fusion reactor, conveniently located some 150 million km from the Earth, radiating 2.3 billion times more energy than what strikes the disk of the Earth, which itself is more energy in a hour than all human civilization directly uses in a year, and it will continue to produce free energy for billions of years.

> "For India to attain at least 90% of the standard of living of one of the developed nations (such as France) a 7% GDP growth rate is essential. Recent studies indicate that as an "insurance policy" to meet potential shortfalls in achieving power capacity growth targets from all terrestrial sources, and for a GDP growth rate of 7%, about 544 GW of solar power from space based stations by 2052 would be required. Such an SSP profile could almost double India's GDP per capita, and delivers a net GDP benefit to the nation of over $100 trillion. The net carbon emission avoided by this SSP growth profile could be about 66 million tons, thus adding to global climate change mitigation efforts."[4]

---

[4] KALAM-NATIONAL SPACE SOCIETY ENERGY TECHNOLOGY UNIVERSAL INITIATIVE, August 16, 2010 , An International Preliminary Feasibility Study on Space Based Solar Power Stations, R. Gopalaswami; India

In space the energy from the Sun is much stronger than on the surface of the Earth because the Earth's atmosphere reflects much of the sun's energy back into space. Solar energy in space is potentially available twenty-four hours a day, every day depending on where and how you try to collect the energy in space. In space, transmission of solar energy is unaffected by the filtering effects of atmospheric gases so you get about 144% of the maximum attainable on Earth's surface. While this might not seem like a huge improvement, this energy is potentially available all the time, so it can really add up.

To capture some of this energy we will need an efficient way to collect the energy in space and transmit it to the surface of the Earth. To do this we will need to build space solar power satellites or PowerSats. These PowerSats can efficiently transmit the energy in the form of microwaves or lasers. In this book we will only consider microwaves however many of the technologies can also be applied to laser based systems.

The entire point of a solar power satellite is to increase the amount of solar energy reaching earth. This extra energy will eventually be dissipated as heat. Depending on the scale of operations, this might or might not have a significant effect. No theories to date claim that waste heat from human power generation are a significant cause of global warming, nor would it be for the foreseeable future.

The mostly widely promoted theory connecting human activity to global warming is that increased greenhouse gases (e.g. Methane and Carbon Dioxide) are causing the natural heat from the Sun to be trapped so it cannot radiate to space, thus increasing the temperature of the planet. Space solar power would contribute greatly to reduction of greenhouse gases.

To the extent mankind's electricity is produced by fossil fuel sources, SBSP offers a capability over time to reduce the rate at which humanity consumes the planet's finite fossil hydrocarbon resources. While presently hard to store, electricity is easy to transport, and is highly efficient in conversion to both mechanical and thermal energy. Except for the aviation transportation infrastructure, virtually all of the World's energy could eventually be delivered and consumed as electricity. Even in ground transportation, a movement toward plug-in hybrids and all electric vehicles would allow a substantial amount of traditional ground transportation to be powered by Space Solar Energy electricity. We would become an Electric Planet.

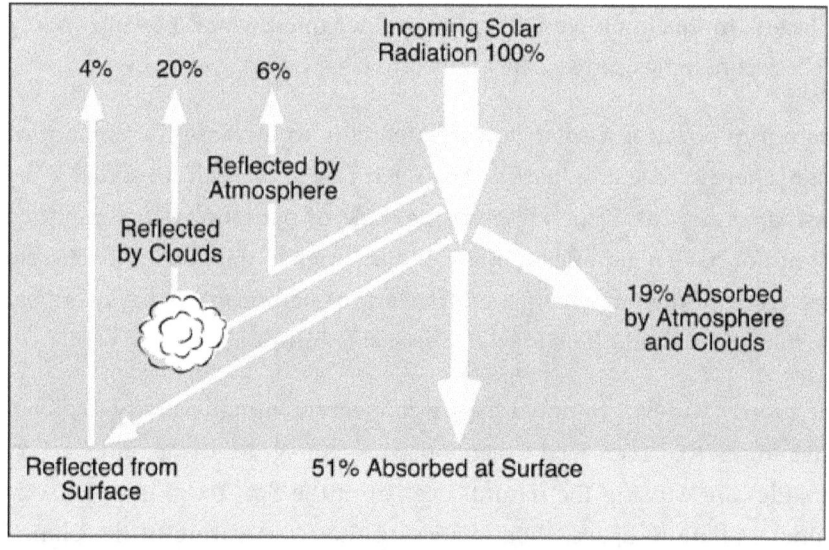

Global modification of incoming solar radiation by atmospheric and surface processes. [5]

---

[5] Image Credit: http://www.solpass.org/6-8Science/6s/standards/study6.3.htm

Scale of Global Energy Use:

"The sun radiates 174,423,000,000,000,000 watts of energy to the earth's cross-sectional area of about 50.3 million square miles or 174,000 Terawatts (TW = 1 Trillion watts). In space above or near the earth, the available solar energy is about 1.3 kw / m2.[2] However, due to the atmosphere, the earth's surface only gets an average of 77 % of the energy available in space or 134,000 Terawatts. The rest is absorbed or reflected by the atmosphere. Only those areas of the earth's surface directly facing the sun (at noon in summer) and without clouds would be getting the maximum of approximately 1.0 kw / m2 of full sun that is possible at the bottom of the atmosphere. All of current human energy production and use amounts to only about 16.4 Terawatts, or about 1 part in 10,000 of the sun's energy hitting the earth. All of earth's wind energy represents about 1.5 % of the sun's energy received by the earth or about 2,600 Terawatts. In contrast, all global photosynthetic plant energy is captured from the sun at a rate of only about 26 Terawatts, less than twice our average global energy needs."[6]

Our ability to implement space solar energy will depend upon making the system much smaller, more compact and lighter weight than previous concepts. Not to worry, we do have the technology to do these things. You might even be amazed at just how well we can do this.

---

[6] J. K. Strickland(2010), Space Solar vs Base Load Ground Solar and Wind Power, Online Journal of Space Communication, Issue 16

Beaming Down:

A space-based solar power station will use an array of mirrors to concentrate the sun's rays on photovoltaic cells. The electricity produced is converted into a powerful microwave beam directed at an antenna on earth, where it is converted back into electricity and fed to the grid.

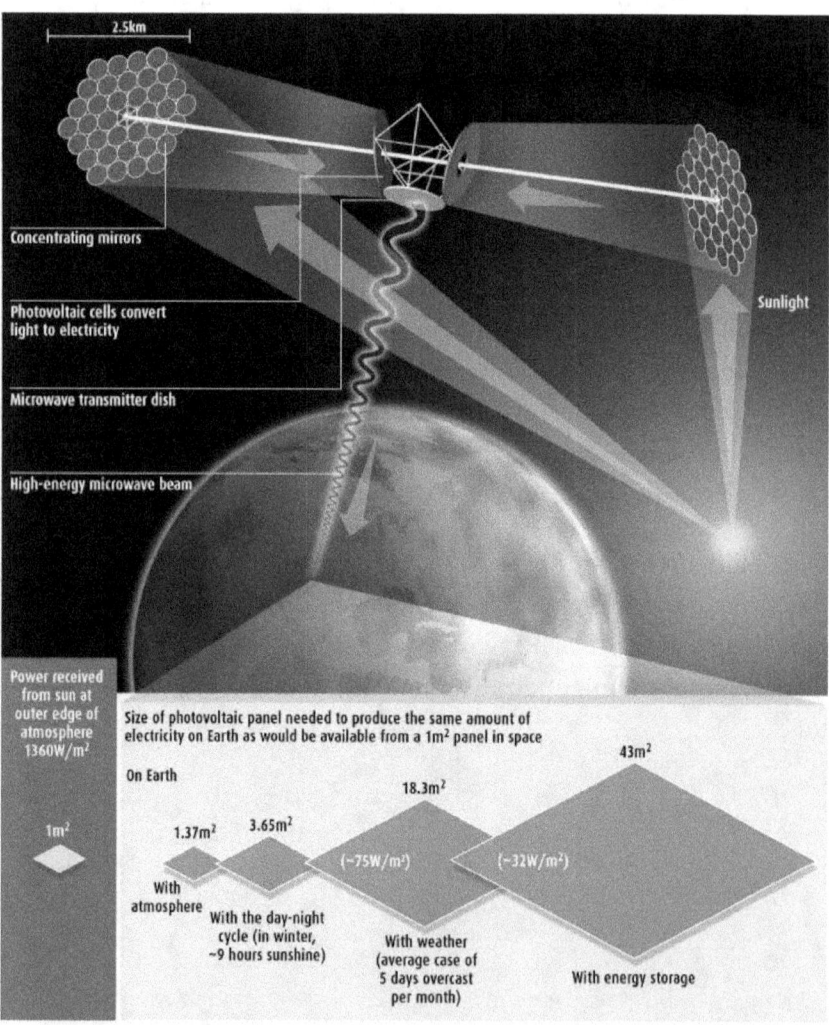

This technology enables a carbon-neutral (closed carbon-cycle) hydrocarbon economy driven by clean renewable sources of power, which can utilize the existing global fuel infrastructure without modification.

This opportunity is of particular interest to traditional oil companies. The ability to use renewable energy to serve as the energy feedstock for existing fuels, in a carbon neutral cycle, is a "total game changer" that deserves significant attention. SBSP's primary environmental benefit is in the form of nearly carbon free, renewable energy. Even considering the energy cost of launch, SBSP systems do payback the energy to construct and launch. In fact, SBSP systems have net energy payback times (<1 year except for very small 0.5 GW plants) well within their multi-decade operational lifetimes. Payback times are equivalent and perhaps faster than terrestrial solar thermal power7. The reason for this is that an equivalent area in space receives 8-10 times the energy flux for the annual average, and as much as 30-40 times the energy flux in a given week than the same area located on a favorable place on the ground after considering day/night, summer/winter, and dust/weather cycles.

Even after losses in wireless power transmission the reduced need for overcapacity and storage to make up for periods of low illumination translates into a much lower land usage vs. terrestrial solar for an equivalent amount of delivered energy. The development of an economically viable space-based solar power (SBSP) system is critical to the Earth's future and for future space development. PowerSat technology is also critical to supporting sustainable private and government space ventures, including space lift, space exploration and space infrastructure development.

Such a system would greatly expand the need for space lift capability from small reusable launch vehicles for SBSP satellite maintenance to large expendable launch vehicles for deploying GW class SBSP satellites into orbit. The technology needed for SBSP is also needed for in-space solar electric transportation systems needed for space colonization as the technology is basically the same.

---

[7] Zerta et al, 2004

While there has been a great deal of debate in the past few years about commercial space transportation into orbit, this book goes beyond space launch and looks at the technologies for sustained commercial development of space both in orbit and beyond using beamed energy. It will consider how many of these technologies can be useful on Earth by providing clean and abundant energy and how people will immigrate into space to live, work and play.

An interesting aspect of these technologies is that they will also spur additional development of our space launch investments and capabilities. This book is about how beamed energy can lead to development of sustainable space markets and how to serve those markets. Space-Based Solar Power taps directly into the largest known energy resource in the solar system – the Sun.

This is not to minimize the difficulties and practicalities of economically developing and utilizing this resource or the tremendous time and effort it would take to do so. Nevertheless, it is important to realize that there is a tremendous reservoir of clean, renewable energy available to the human civilization if we can develop the means to effectively capture it and move it around. It is not really surprising that Space Solar Power (SSP) has not been economically viable given the huge mass to orbit requirements of Space Solar Power Satellites (PowerSats).

In order for SSP to be economically viable massive reductions in satellite mass are necessary. Fortunately, there are ways to reduce the mass of PowerSats. Several mass reduction strategies will be discussed, such as Massive Solar Concentration (MSC), Active Photovoltaic (PV) cooling alternatives, Reflector Technology, higher frequencies and beam densities and some Alternative orbits that place the PowerSat closer to Earth to reduce transmitter mass.

The original SSP base-line models were not only hugely uneconomical they were also technically impossible. The authors say technically impossible because they required massive reusable launch vehicles (RLVs) that are none existent and would be difficult and expensive to build and also required thousands of workers transporting to and living in space for decades to build an SSP satellite in geostationary orbit (GEO). We need to be able to start out smaller and more economically.

There is a scenario that allows this to happen and that scenario would be to use power beaming to support an in-space transportation system. The technology necessary to do this is essentially the same as that required for SSP. By using SSP type technology for space transportation we can deploy much smaller PowerSats that can be used to substantially lower the cost of space transportation while at the same time serving as precursors to larger PowerSats for Earth energy.

Chapter Three

Space-based Solar Power Satellites

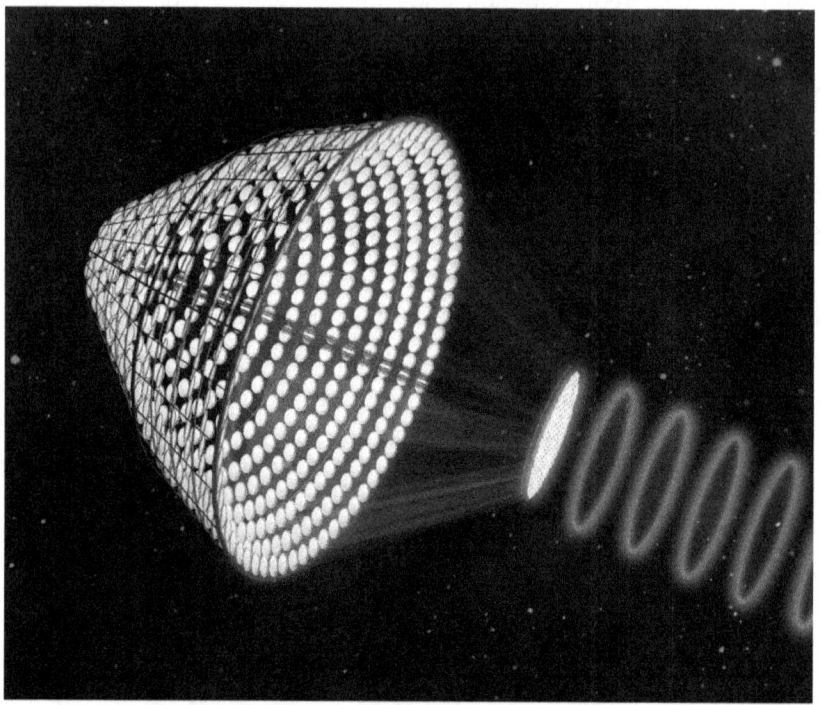

SPS-ALPHA (Solar Power Satellite via Arbitrarily Large Phased Array) - is a novel, bio-mimetic approach to the challenge of space solar power[8]

The solar energy collected by an SPS would be converted into electricity, then into microwaves. The microwaves would be beamed to the Earth's surface, where they would be received and converted back into electricity by a large array of devices known as a rectifying antenna, or rectenna. Rectification is the process by which alternating electrical current, such as that induced by a microwave beam, is converted to direct current.

---

[8] Image retrieved from http://www.nss.org/settlement/ssp/library/SPS_Alpha_2012_Mankins.pdf

This direct current can then be converted to the "slower" 50 or 60 cycle alternating current that is used by homes, offices, and factories. In the past a solar power satellite (PowerSat) was a very large-area satellite in an appropriate orbit, which would function as an electric power plant in space. The satellite would consist of three main parts: a solar-energy collector to convert solar energy into dc electric power; a dc-to-microwave converter; and a large antenna array, to beam the microwave power to the ground. Solar Power Satellites or PowerSats have been proposed as huge satellites designed to collect space solar energy and beam the energy to Earth. The 1979 SPS designs consisted of large, erected infrastructures.

These massive units required a very large two-stage Earth-to-orbit (ETO) transportation system to lift the needed material as well as a massive construction facility in space and hundreds of astronauts. The financial impact of this deployment scheme was significant. In 1986 dollars, more than $250 billion was estimated to be required before the first commercial kilowatt-hour could be delivered. (Mankins, 1997) Today's cost would of course be much higher if we tried to build those designs. However, improvements in technology over the last four decades would allow us to build smaller, more compact designs which would have considerably less mass. This would make it easier to build the PowerSats and launch them into space.

Most early (1979), more popularized, PowerSat concepts were based on very large collectors and transmitters in distant geostationary orbit (GEO) and required a large astronaut construction force to construct and maintain them. Such GEO concepts were also based on heavy lift launch capabilities that had not, and have not yet been developed. These popular though unviable, concepts assumed that each PowerSat would be required to be stationary over their respective ground receiving stations and thus were placed in GEO orbit.

However, the disadvantage of geostationary orbits is the great distance over which the PowerSat must transmit power. At the distance of geostationary orbit, ~35,786 km (22,236 mi) (and directly above the equator) a PowerSat transmitter would, necessarily, be very massive. They depended on the use of wireless power transmission using microwave to move the energy from space to the surface of the Earth.

For the production of 5 GW of dc power, the solar collector would need to have an area of 10 square kilometers, and would consist of either photovoltaic cells or solar thermal turbines. The dc-to-microwave converter could be realized using either a microwave-tube system or a semiconductor system, or a combination of both. For transmitting the power to the ground, frequency bands around 2.45 Gigahertz (GHz) or 5.8 GHz have been proposed, which are within the microwave radio windows of the atmosphere.

At 2.45 GHz the transmitter would have a mass of about 13 million kilograms and at 5.8 GHz the mass would be about half or 6.5 million kilograms. The antenna array to transmit the energy to the ground would require a diameter of about two kilometers at 2.45 GHz, and its beam direction would have to be controlled to an accuracy of significantly better than 300 meters on the Earth, corresponding to 0.0005°, or less than 2 arc seconds (for a geostationary orbit of the satellite).

At 5.8 Ghz the transmitter would be about half the size. At even higher frequencies, such as 15 Ghz the transmitter would be even smaller. However, at higher frequencies there would be additional energy losses so a trade study would need to be done to determine the mass verse cost at different frequencies.

Advantages of Space Solar Power:

• Unlike oil, gas, ethanol, and coal plants, space solar power does not emit greenhouse gases.

• Unlike coal and nuclear plants, space solar power does not compete for or depend upon increasingly scarce fresh water resources.

• Unlike bio-ethanol or bio-diesel, space solar power does not compete for increasingly valuable farm land or depend on natural-gas-derived fertilizer. Food can continue to be a major export instead of a fuel provider.

• Unlike nuclear power plants, space solar power will not produce hazardous waste, which needs to be stored and guarded for hundreds of years.

• Unlike terrestrial solar and wind power plants, space solar power is available 24 hours a day, 7 days a week, in huge quantities. It works regardless of cloud cover, daylight, or wind speed.

• Unlike nuclear power plants, space solar power does not provide easy targets for terrorists.

• Unlike coal and nuclear fuels, space solar power does not require environmentally problematic mining operations.

• Space solar power will provide true energy independence for the nations that develop it, eliminating a major source of national competition for limited Earth-based energy resources.

• Space solar power will not require dependence on unstable or hostile foreign oil providers to meet energy needs, enabling us to expend resources in other ways.

• Space solar power can be exported to virtually any place in the world, and its energy can be converted for local needs — such as manufacture of methanol for use in places like rural India where there are no electric power grids. Space solar power can also be used for desalination of sea water.

- Space solar power can take advantage of our current and historic investment in aerospace expertise to expand employment opportunities in solving the difficult problems of energy security and climate change.

- Space solar power can provide a market large enough to develop the low-cost space transportation system that is required for its deployment. This, in turn, will also bring the resources of the solar system within economic reach.

Disadvantages of Space Solar Power:

High development cost. Yes, space solar power development costs will be very large, although much smaller than American military presence in the Persian Gulf or the costs of global warming, climate change, or carbon sequestration. The cost of space solar power development always needs to be compared to the cost of not developing space solar power.[9]

Before we jump into this subject we should take a few minutes to consider some of the basics concepts such as Wireless Power Transmission, Microwave beaming and the effects of a beam spreading over distance. While the use of laser transmission might be useful in this book we are only going to consider microwave power beaming because of its high efficiency. W do encourage others to take a look a laser transmission and apply some of the technologies and concepts contained herein to the laser based systems.

---

[9] J C Mankins, 1997, "A Fresh Look at Space Solar Power: New Architectures, Concepts and Technologies", IAF paper no IAF-97-R.2.03, 38th International Astronautical Congress

Wireless power Transmission:

Wireless communication uses radio waves as carriers of information. However, in the microwave power-transmission system, radio waves would be used as carriers of energy. In principle, the energy-carrying microwaves would be monochromatic waves, without any modulation. The microwave power transmission would use power densities at the surface of the transmitting antenna that are three or four orders of magnitude higher than the corresponding levels in wireless-communication systems, and up to 25 orders of magnitude higher than power densities received by the radio-astronomy and remote-sensing services.

The main parameters of the microwave power-transmission system for the SPS system are the frequency, the diameter of the transmitting antenna, the output power beamed to the Earth, and the maximum power-flux density. In addition to the system parameters described above, the weight per unit power of the microwave devices is also of importance. Efficiency is very important for the microwave power-transmission system. Assuming the SPS transmitting-antenna-to-rectenna propagation path is optimum, the following efficiencies will be important: dc-to-radio-frequency (RF) conversion, RF-to-dc conversion, and beam-collecting efficiencies. Conversion efficiencies higher than 80% for both RF-dc and dc-RF conversions are necessary to make the cost of the SPS system reasonable.

Various types of transmitting antennas have been considered, such as slotted-waveguide antennas, dipole antennas with reflectors, and microstrip antennas. The most suitable antenna type depends on the chosen microwave generator and amplifier, but also on weight. A possible concept seems to be the active integrated antenna technique, combing the dc power generation, microwave conversion, and radiation and control in one multi-layered plate. However, conventional tube transmitters can offer very high beam levels.

Wireless power Transmission (WPT) and it relation to space may be thought of as extending our two dimensional power transmission networks on the Earth to space and to other planets and space vehicles. Such a system could be used for a wide variety of applications.

One such application would be providing large amounts of power for an electric spaceship needed for an in-space transportation system traveling back and forth to the Moon and Mars. Electric propulsion has long been recognized for its benefits if there were a suitable energy source for the large amounts of power required by electric thrusters.

Conventional prime power sources in space are massive relative to electric thrusters and must be accelerated along with the less massive parts of the vehicle. Further, they are expensive and costly to transport into space. In contrast, beamed microwave power removes the prime power source from the vehicle and therefore has a very low mass relative to other potential prime power sources in space, including chemical, nuclear and solar electric.

Mass savings can range from 20 to 30 percent. The combination of WPT and electric thruster technology would make it possible to replace conventional chemical rocket propulsion for missions beyond low Earth orbit (LEO) with enormous economic benefits. By pursuing Space Solar Energy to supply the Earth with energy we are also developing the technology for large scale colonization of the solar system at the same time. In fact, given the huge cost of space transportation we would want to build space to space beamed energy systems first. These would be precursors to a Space to Earth energy system.

Microwave Power Transmission:

Microwave power transmission (MPT) involves the usage of microwaves to transmit power through outer space or the atmosphere without the need for wires. It is a sub-type of the more general wireless energy transfer methods, and is the most interesting because microwave devices offer the highest efficiency of conversion between DC-electricity and microwave radiative power. One of the biggest potential applications of microwave power transmission is its utility in solar power satellite systems, or SPS.

Following World War II, which saw the development of high-power microwave emitters known as cavity magnetrons, the idea of using microwaves to transmit power was researched. In 1964, William C. Brown demonstrated a miniature helicopter equipped with a combination antenna and rectifier device called a rectenna.

The rectenna converted microwave power into electricity, allowing the helicopter to fly. In principle, the rectenna is capable of very high conversion efficiencies - over 90% in optimal circumstances.

> "The Industrial-Scientific-Medical (ISM) frequency bands available for microwave transmission and accepted by international regulatory authorities include the 2.45-, 5.8-, and 24.125-GHz bands. The US DOE, Raytheon Co., NASA's Jet Propulsion Laboratory (JPL), the Canadian Communications Research Center, and various European agencies have taken part in development programs for the components needed for microwave power transmission at 2.45 GHz, due to the low atmospheric attenuation at that frequency. The 5.8-GHz band is preferred by a number of research organizations, including the David Sarnoff Research Center, SRI International, NASA, and JAXA, because of the reduction in component size at that smaller wavelength compared to 2.45 GHz.

In addition, the attenuation loss through the atmosphere does not increase appreciably at 5.8 GHz compared to 2.45 GHz."[10]

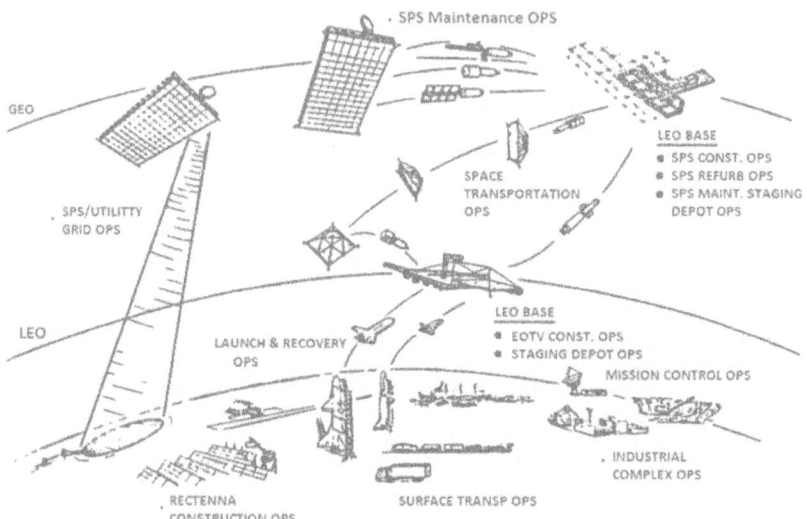

The past Infrastructure rich Space Solar Energy concepts required massive investments to carry crews and cargo into space to assemble giant SSP satellites in GEO. Source: Integrated Space Operations Overview, Gordon R. Woodcock, Boeing Aerospace Co. 1980

The Rectenna is a key component in a power transmission system since it must handle high power levels with high efficiency in order to conserve as much as the received energy from space as possible. Early work on a thin-film rectenna at 2.45 GHz was reported by Brown. The rectenna achieved conversion efficiency of 85 percent at S-band, reported a highly efficient antenna array at 5.87 GHz, comprised of 1000 dipole elements. These rectenna dipole elements exhibit RF-to-DC efficiency exceeding 80 percent with a uniform illuminated aperture. The rectification element consists of a silicon Schottky diode quad bridge with high reverse breakdown voltage.

---

[10] Kumar, A (May 2010), Antenna assists MW power transmission: this rectenna dipole element uses matching stubs for increased power capability and improved efficiency for microwave-power-transmission applications at 5.8 GHz. Microvaves & RF, p133

The prevailing popularity, though economically and technologically non-viable, GEO Space Solar Energy concepts requiring a large astronaut work force and non-existent heavy-lift vehicles has arguably detracted from the voices that have, from the outset, called for the development of economically viable concepts that place smaller Space Solar Energy systems in Low to Middle Earth orbit. Such ideas have been largely pushed aside in favor of the GEO location.

Such GEO based approaches are of course counterproductive because economically unviable concepts actually limit progress not only for power production but also in development of new launch vehicles and space infrastructures. While a few investigators were discussing LEO and MEO SBSP satellites as far back as the 1970s, their ideas have been largely pushed aside in favor of the GEO location. (Drummond, 1980) This illogical addiction to an unnecessary GEO infrastructure is the principal reason so little progress has been made in SBSP. The first step in demonstrating the economic viability of Space Solar Energy is to move away from past concepts based on GEO solar power satellites and a large astronaut workforce.

Radiation and the Inverse Square Law:

The rate at which energy emanating from a fixed, constant source of electromagnetic radiation passes through a surface at a distance d from the source is proportional to $1/d^2$. This is known as the Inverse Square Law. It arises simply because the surface enclosing the source is a sphere, centered on the source, through which all the energy must pass and the surface area of this sphere increases as the square of the distance d from the source. Thus the energy flow (measured in Watts per square meter (W/m^2)) falls off rapidly as the distance from the source increases.[11]

---

[11] Antenna assists MW power transmission: this rectenna dipole element uses matching stubs for increased power capability and improved efficiency for microwave-power-transmission applications at 5.8 GHz.(MW POWER TRANSMISSION) Article from: Microwaves & RF | May 1, 2010 | Kumar, A.

The Physics of Wireless Power WPT Beam Intensity Relationship

$$\text{Rcvr Peak Intensity}_{Beam} = \pi \times \frac{\text{Power (Xmitter)}}{8} \times \left[\frac{\text{Diameter (Xmitter)}}{\lambda_{Beam} \times \text{Sep'n}_{Xmitter\ to\ Rcvr}} \right]$$

Space-Based Solar Power as an Opportunity for Strategic Security, Phase 0 Architecture Feasibility Study Report to the National Security Space Office October 10, 2007, Page 27

The intensity (or illuminance or irradiance) of light or other linear waves radiating from a point source (energy per unit of area perpendicular to the source) is inversely proportional to the square of the distance from the source; so an object (of the same size) twice as far away, receives only one-quarter the energy (in the same time period). More generally, the irradiance, i.e., the intensity (or power per unit area in the direction of propagation), of a spherical wavefront varies inversely with the square of the distance from the source (assuming there are no losses caused by absorption or scattering).

The fractional reduction in electromagnetic fluence (F) for indirectly ionizing radiation with increasing distance from a point source can be calculated using the inverse-square law. Since emissions from a point source have radial directions, they intercept at a perpendicular incidence. The area of such a shell is where r is the radial distance from the center.

What this means in plain English is that the greater the distance the greater the beam spread. This means larger transmitters and larger receivers (rectenna) are needed as the distance increases. For example, the intensity of radiation from the Sun is 9126 watts per square meter at the distance of Mercury (0.387 AU); but only 1367 watts per square meter at the distance of Earth (1 AU)—an approximate threefold increase in distance results in an approximate nine-fold decrease in intensity of radiation. Therefore, a PowerSat beaming to a distance of 12,000km would put the same energy onto a "spot" nine times smaller than if beamed from GEO at 36,000km.

36,000km/12,000km = 3 squared = 9

6,500,000kg transmitter mass / 9 = 722,222kg

The satellite spot beam could be over 9 times smaller if it beamed from a 12,000km orbit rather than GEO, which is 36,000km. This means each square meter of area on the ground will receive nine times as much energy. If we wanted to the keep the energy level the same we could use nine smaller satellites. Therefore, we now have a scenario which would allow the deployment of much smaller PowerSats rather than trying to build giant PowerSats in GEO.

The question now becomes very simple - Does the reduced mass of moving the PowerSat closer make up for the increased mass needed to offset Earth shadowing? This is a question we will explore in great detail throughout this book.

Chapter Four

Brief History of the Solar Power Satellite Concept

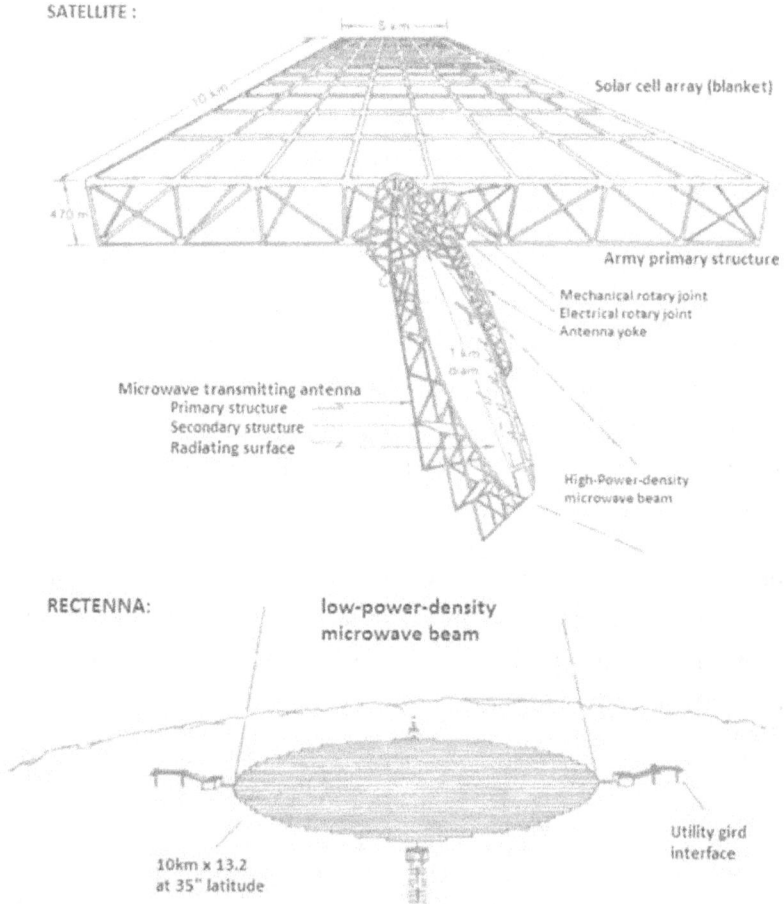

Electric Power from Orbit: A Critique of a Satellite Power System. National Research Council of the National Academy of Sciences, 1981

---

[12] Image Adopted from U.S. DOE(1980d)

The PowerSat designs that became popular were those that envisioned the PowerSat in geostationary orbit as large panels or towers of photovoltaic (PV) solar energy collectors that would convert incident sunlight directly into electricity that would in turn, be transmitted via microwave to ground receivers. These massive units were designed to require a two-stage Earth-to-Orbit (ETO) space transportation system to lift the needed materials, and hundreds of astronauts to work from a construction facility in space for years. The financial impact of this deployment scheme was significant. In 1986 dollars, more than $250 billion was estimated to be required before the first commercial kilowatt-hour could be delivered. The dimensions of the NASA baseline Space Solar Energy concept from 1979 are shown below. The concept has a system mass of approximately 51,000,000 metric tons.

The U.S. National Research Council (NRC) and the Congressional Office of Technology Assessment (OTA) concluded that solar power satellites were technically feasible, but were declared "programmatically and economically unachievable" based on the 1979 SPS Reference designs.

Although the NRC recommended that related research continue and that the issue of solar power satellite viability should be revisited in about a decade, in fact all serious effort on solar power from space by the U.S. government ceased.

The NRC report stated,
> "Too little is currently known about the technical, economic, and environmental aspects of SPS to make a sound decision whether to proceed with its development and deployment. In addition, without further research an SPS demonstration or systems-engineering verification program would be a high-risk venture."

I find it interesting that proponents of SSP are still pushing this same concept even though technology has already made such a system obsolete.

Between 1995-1997, the National Aeronautics and Space Administration (NASA) re-examined the technologies, systems concepts and terrestrial markets that might be involved in future space solar power systems during 1995-1997.

Its principal objective was to determine whether solar power satellites could deliver energy "to terrestrial electrical power grids at prices equal to or below ground alternatives in a variety of markets, do so without major environmental drawbacks, and which could be developed at a fraction of the initial investment projected for the Reference System of the late 1970s. Out of this NASA re-examination, three potential architectures were identified; a sun-synchronous Low Earth Orbit (LEO) constellation, a Middle Earth Orbit (MEO) multiple-inclination constellation, and one or more stand-alone Geostationary Earth Orbit (GEO) systems. Of particular interest was the Sun Tower concept because it offered a much smaller transmitter size hosted in a closer orbit. So let's take a look at that concept.

The Sun Tower:

Since the sun provides about 1365 watts per square meter of energy at the Earth's orbit, generating a megawatt with a 20% efficient array requires an area of about 3700 square meters. However, the SPS concept that emerged by 1979 was not only large, it was also infrastructure-rich because it was based upon the large, astronaut-erected space platform concepts that were common of this era in which Gerard O'Neil and others envisioned the eventual construction of vast, artificial cities in space.

As microwave transmitter size and mass is a direct function of distance between transmitter and receiver, by bringing the PowerSat closer to Earth, a smaller transmitter could be used to provide the same level of power as one that must transmit energy over a further distance. Simply by moving the SBSP system closer to Earth, both the mass the transmitter and the size of the ground receiver could be substantially reduced and bring the concept of Space Solar Energy closer to economic viability. This was the original idea behind the Sun Tower.

Given the physics of wireless power transmission, when compared to geosynchronous (GEO) orbit at 36,000 km, medium (MEO) Earth orbits located at 12,000 km or less, should permit considerable reductions in the size of both the solar power transmitter and the ground receiver. Furthermore, a smaller ground receiver is better suited to servicing such high-density markets as exist in Japan and Western Europe.

The Sun Tower concept of the 1995-1997 NASA SBSP study, with its constellation of solar collectors and a large transmission dishes pointing Earthward, was envisioned as being placed in a closer, MEO, sun-synchronous, 1,100 km altitude orbit rather than the distant GEO orbit and therefore required a much less massive space based system than those proposed in the late 1970s. The ground area covered by the Sun Tower's receivers was also dramatically smaller than that required for earlier, GEO based SBSP designs.

The space based segment of the Sun Tower concept was a constellation of medium-scale, gravity gradient-stabilized, satellite towers resembling a large, Earth-pointing sunflower with its RF-transmitter array being the flower and leaves of PV solar collectors reach from the stalk of the tower. The Sun Tower's transmitter arrays were envisioned to be 260 meters in diameter and up to a meter thick.

It is easy to see that the Sun Tower concept located in MEO is a major improvement over a GEO location due to its much lower mass that needed to be delivered to orbit. Even so, papers are still published that are based on GEO concepts with high mass requirements.

There are even papers that move the MEO Sun Tower to GEO. Most such papers come from the U.S. aerospace community which seems to be driven more by the desire to build unnecessary infrastructure that requires building massive new launch vehicles than by the desire to produce power in space. While PowerSats would be located in space they are really and Energy Project and should be considered from that prospective.

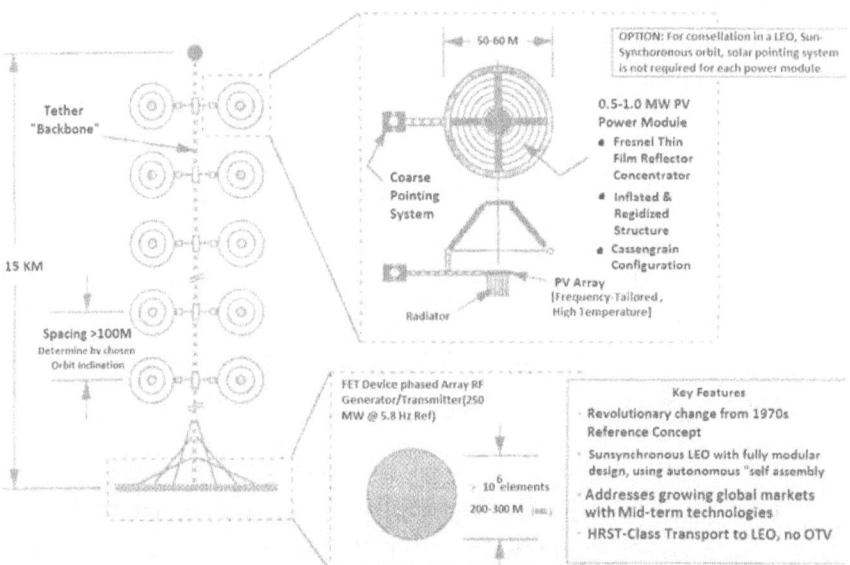

The "Sun Tower" SPS Concept (MEO constellation), Source: A Fresh Look at Space Solar Power: New Architectures, Concepts and Technologies, page 11

By moving the PowerSat constellation from GEO to MEO, the Sun Tower concept represented a major improvement over earlier GEO based designs. The Sun Tower was a dramatically less costly approach than the previous PowerSat designs based in geosynchronous orbit because the size and mass of the transmitter could be substantially reduce and the space transportation requirement was much less. However, the Sun Tower still required a massive satellite system, a large astronaut force and heavy lift vehicles yet to be developed.

The Sun Tower also had its own set of problems, such as Earth shadowing, limited beam time and the large size of the Earth's oceans. The Sun Tower was therefore just as economically and programmatically unviable as the earlier, more distant GEO based SBSP concepts. The Sun Tower, like earlier GEO based SBSP systems, being economically, politically and technologically infeasible, was shelved.

Sun Tower in LEO 1,000km:

Drummond of Power Conversion Technology, Inc. calculated that smaller power blocks will increase market penetration by opening smaller markets (including those in the Third World), by lowering costs of service to decentralized markets, and by smoothing introduction of the SPS power into the Grid." (Drummond, 1980) What is being proposed is the need to break up the satellite system into more economically affordable systems. By having several smaller SBSP satellites operating in network, the system can be deployed incrementally and is therefore more affordable than building a single giant satellite.The ability of space solar providers to begin delivering power early in the constellation deployment and to be able to incrementally increase constellation size can add to the affordability factor.

Following this model, space energy providers don't need to spend hundreds of billions of dollars to build a single massive satellite when smaller systems will serve the purpose.

> "For 200 MW transmitted RF power, the transmitter array is approximately 260 meters in total diameter, and approximately 0.5-to-1.0 meters in thickness." [13]

Sun Tower in MEO :

The Middle Earth Orbit (MEO) Sun Tower, consisted of a 15-km gravity gradient backbone with 340 pairs of solar collectors. At the bottom of the backbone was a circular 300-m nadir-pointing phased array transmitter that would beam power to the Earth at a frequency of 5.8 GHz. The satellite would be in a circular equatorial 12,000-km orbit.

Note the transmitter sizes are 260-m for the 1,100km tower and 300-m for the 12,000 km tower.

It is easy to see that the Sun Tower concept located in MEO is a major improvement over a GEO location due to its much lower mass to orbit. Even so, papers are still published that are based on GEO concepts with high mass requirements.

There are even papers that move the MEO Sun Tower to GEO. Most such papers come from the U.S. aerospace community which seems to be driven more by the desire to expand man's domain into space that requires building massive new launch vehicles than by the desire to produce power in space.

---

[13] John C Mankins, A Fresh Look at Space Solar Power: New Architectures (1997), Concepts and Technologies, NASA, DC, USA

Chapter Five

Technical Problems

Two different PowerSat concepts from the past.

There are a number of problems related to PowerSat deployment and operations. Some of these problems are compounded by PowerSat orbital location. The Main shadowing Problem is LEO satellites that much of the orbit time is spent in the Earth's shadow and no solar power can be produced. In this Chapter we are going to Review such technical problems.

Shadowing Problem:

For LEO satellites much of the orbit is spent in the Earth's shadow and no solar power can be produced. This means you have to launch even more satellites to make up for the estimated 40% power loss.

The period of an orbit is; T = 2 * pi * SQRT(a^3 / mu)

mu = 398,600.4418 km3/s2

T = 90 minutes = 5,400 seconds

Rearranging the equation a = mu (T/(2 * pi)^2)^(1/3)

Obtains an altitude of 286.36 km - and the utilization is precisely 59.37%

As the PowerSat is moved out in distance the shadow time becomes less. At 12,000 kilometers the shadow time is about 12 percent and at GEO the shadow time is less than 1 percent. There is however a way to solve the shadow problem.

This can be achieved by placing the PowerSat is a sun-sync orbit and using reflectors in equatorial orbit to move the power around. I invented this concept, which I call The Space Grid in 2010. Additional detail on how it works will be discussed later.

Beam Time Problem:

Because the Earth rotates under the satellite the beam time to a ground rectenna would be very short at roughly 200 -300 seconds for an LEO PowerSat (Jones R. (3)). Since the SBSP satellites would be very close to the Earth and would be traveling very fast there is little time to beam to power to the rectenna. This problem can also be solved using a space power relay (SPR) with reflectors.

By placing approximately ten reflectors in a 4,000km equatorial orbit we can achieve constant power as shown below.
Time over target 18 minutes
4,000 km orbit = 175.32 sec orbit / 18 minutes of beam time = 9.74 reflectors (rounded to 10)

Ocean Problem:

Another problem with LEO PowerSats is the Earth's Oceans. For example, the distance from Indonesia to the coast of Colombia and Peru across the Pacific Ocean is 19,800 kilometers. The Space Power Relay (SPR) reflectors can help overcome this problem. There are some islands or groups of islands in the Pacific that could use space solar power, one of those is called Hawaii. The SPR reflectors can overcome this problem because the PowerSats can choose which reflector to use to beam to the Earth. By having reflectors in space it would be possible to move the power to where it is most needed.

The benefits of SPS deployments in LEO/MEO impact not just the space segment, i.e. the space transmitter, but also the ground receiver. According to Kotin writing in 1978, the total land area required by each rectenna facility, including provision for a microwave buffer zone, based on GEO-located satellites is estimated at approximately 50,000 acres or 200 square kilometers. By locating the satellites in LEO ground receiver size is reduced by over 90%. Past cost estimates for ground systems using the GEO satellite reference exceed $2 billion. Alternately, LEO satellite system ground receivers using the Sunflower concept will require more or less 4 square kilometers of space, costing a small fraction of the GEO system ground receiver.

A new MEO PowerSat design was proposed by the author in 2010. This new concept placed the PowerSat in a near polar sun-synchronous orbit at 800km and used microwave reflectors (space power relay) to bounce the microwave beam to the ground.

This solved the problems of shadowing, beaming time and getting around the oceans. This concept called The Space Grid solved the problems that plagued the Sun Tower and other LEO concepts. In 2011 this concept was modified by placing the PowerSat in a 2,722km sun-sync orbit at 110 degrees. This moved the PowerSat closer to the space reflectors and shorted the beam distance.

Space Launch Problem:

There are a number of options available to increase the cost effectiveness of PowerSats. First and not surprising at all, is that launch costs made up a really big portion of the total project cost of past PowerSat concepts so the primary focus has to be on reducing the mass of the PowerSat so that they can be launched cost effectively. Second, there is a demonstrated trend in using higher solar concentrations as a way to reduce the mass of the SSP PowerSat. Third there are options other than GEO which needed to be looked at much closer, because moving the satellite closer could reduce the size and mass of the satellite power transmitter and these options have not been fully investigated. Forth, increasing the beam power, even at higher atmospheric power loses, could also reduce transmitter size and mass and last, much of the proposed space infrastructure could be eliminated if the satellite could be launched as a complete unit without assembly in orbit.

The one of primary challenges for SSP satellites is the cost of space launch. Over half the cost of past SBSP system designs is associated with launch costs. To reduce launch costs the size and mass of the SSP system must be reduced. In order for SSP concepts to become an economically viable source of clean energy, it is necessary to lower the SPS satellite's mass and size to the point that it can be launched into working orbit with currently available or very near term commercial launch vehicles – and without requiring any on-orbit astronaut assembly or unnecessary infrastructure.

So how can we improve the economical viability of SSP PowerSats?

The first objective is to reduce the mass of the PowerSat and this will greatly affect the type of launch vehicle needed. Most all literature in the last forty years has tried to solve the PowerSat mass problem by proposing very large, hugely expensive and often fantasy rockets to put the mass into orbit. Here we are going to take a completely different approach and focus on mass reduction as the primary solution.

The first thing we would want to do is look closely at the PowerSat mass problem. We can see that there are two major systems components, the power generating system and the microwave (or laser) transmitter. The mass of the power generating system depends both of the amount of power you want to generate and the technology used to generate it. The transmitter mass is determine mostly by the distance the energy is transmitted.

The Mass Problem:

SBSP is a very interesting subject. You already know that SBSP is not economically viable as presented in the literature. All of the reports and books say the same thing, which is, for SBSP to be economically viable you need massive and unrealistic reductions in launch costs. So, this is a mass to orbit problem. The hope has been that gradual improvement in Photovoltaic (or other technologies such as thermal) efficiency will solve the mass problem. However, increased efficiency in itself is not sufficient to make SBSP economically viable.

The popular interest in developing a large astronaut force and new heavy lift launch vehicles to develop a GEO SBSP system seems to have overwhelmed the voices of those who, since the early 1970's, have suggested that by simply moving the satellites closer to the Earth, both the transmitter and the receiver can be greatly reduced in size and mass, and thus could significantly reduce the cost of the system and make the concept of SBSP economically, technically and politically viable.

Transmitter Mass Problem:

The largest potential application for microwave power transmission is SBPS satellites. In this application, solar power is captured in space and converted into electricity and beamed to the Earth. Several concepts have been proposed in the past for LEO PowerSat beaming to Earth to alleviate the launch cost problem (2, 9). It has been known since at least 1980 that placing PowerSats in LEO would reduce satellite transmitter mass by "an order of magnitude" (Drummond (2)), i.e., about 90%. However, the problems of PowerSat stationing in LEO are the Earth's rotation under the satellite, Earth shadowing and the Oceans. A comparison of the NASA/DOE SBSP concepts dating back to the 1980s shows the mass problem related to SBSP satellites in Geostationary orbit (GEO) (12,13,14,15,16). The transmitter mass problem for GEO PowerSats can be seen below.

NASA 1980 Option 1: 1x Concentration, 16% efficient PV, 5GW, Mass 51,000,000kg, Transmitter 13,000,000kg, Power 38,000,000kg

NASA 1980 Option 2: 2x Concentration, 20% efficient PV, 5GW, Mass 34,000,000kg, Transmitter 13,000,000kg, Power 21,000,000kg

ISC 1990: 4x Concentration, 20% efficient PV, 5GW, Mass 23,500,000kg, Transmitter 13,000,000kg Power 10,500,000kg

ISC 2010: 4x Concentration, 40% efficient PV, 5GW, Mass 18,250,000kg, Transmitter 13,000,000kg, Power 5,250,000kg

Notice above that even when you double the efficiency of the power system by doubling the solar concentration or doubling the photovoltaic (PV) efficiency or both that the transmitter mass is still the same at 13 million kilograms for a 5GW transmitter. It becomes very clear that new approaches to mass reduction are needed for SBSP and BSEP to become economically viable.

One way to reduce the mass is to move the transmitter closer but there are some problems in doing that. These problems include reduced beam time due to satellite speed around the Earth and the Earth's rotation, Earth shadowing which blocks the satellite for the suns energy and the large size of the oceans that make it difficult to transmit energy when you are very close to the Earth.

One of the WPT proposal's disadvantages is that microwaves have long wavelengths that exhibit a moderate amount of diffraction over long distances. The Rayleigh criterion dictates that any beam (microwave or laser) will spread, become weaker, and diffuse over distance. This is just like pointing a flashlight at a wall and then walking backwards. The larger the transmitter antenna or laser aperture, the tighter the beam and the less it will spread as a function of distance (and vice versa). Therefore, the system requires large transmitters and receivers if located far away from each other. We could reduce the transmitter mass by half by moving to 5.8Ghz

ISC 2010: 5.8Ghz 4x Concentration, 40% efficient PV, 5GW, Mass 11,750,000kg, Transmitter 6,500,000kg, Power 5,250,000kg

Again we see that transmitter mass is the same in all cases (although total satellite mass is less). So, we have already identified two important trends, higher levels of solar concentration and higher beam frequencies.

Power Production Problem:

The SPS satellite concept can be regarded as quite elegant: a large platform, positioned in space in Earth orbit continuously collects and converts solar energy into electricity. This power system then uses a wireless power transmission (WPT) system (very similar to the computer modem you may have in your home) that transmits the solar energy to receivers on Earth. Because of its immunity to nighttime, to weather or to the changing seasons, the SPS concept has the potential to achieve much greater energy-efficiency than ground based solar power systems. Most studies of Space Solar Power (SSP) to date have assumed photovoltaic (PV) solar energy conversion and satellite stationing in geostationary orbit (GEO). The problem with such concepts is that they are simply too massive. To solve this problem we need to consider the current state-of-the-art for photovoltaic and related technologies, such as using solar concentration and spectrum beam splitting.

The GEO Problem:

"Thirty six thousand kilometers above earth is a logical destination for a number of reasons, but that orbit is already largely committed. What is more, this great height and the mass and number of space solar systems proposed for GEO will not be cost-justifiable anytime soon. Decades will pass before this promising location will be a major solar power satellite (SPS) destination due to incumbent player resistance over possible signal interference. Also, dramatic improvements in space-based PV cell technology will be needed, as will reductions in the cost of space launch. SPS systems will be a predictable contributor to our energy future when these birds are built to operate in space at costs competitive with energy systems on Earth. Successful SPS designs will be those that are technically feasible, economically affordable and can be proven to work.

Most solar power system placement proposals are intended for geosynchronous orbit. This is one reason the GEO solar power satellite (SPS) systems end up having an initial start up cost of hundreds of billions of dollars. The largest single cost of GEO solar power satellites is the cost of launching the components into orbit. The second largest cost is moving the components from low Earth orbit (LEO) to geostationary (GEO).

The only real benefit of GEO is the ability to stay over the target receiver constantly. The major negative points of GEO are 1) Transmitter size – this is a major issue and possibly the killer for GEO concepts. 2) LEO to GEO transfer – the cost of this transportation is very substantial, past concepts including ion propulsion, which is the best approach still have huge costs in propellant mass to orbit. 3) GEO is already highly populated with ComSats – even if you could solve the other problems this would still kill any GEO SBSP satellite proposal.

The main problem with GEO SPS is the 36,000 kilometer distance. This distance from Earth requires large microwave transmitters and large ground receivers. The great distance also results in very high launch costs due to the transmitter size and mass and the very real prospect of interference with the large number of communication satellites located there.

The reason that the solar power satellite must be so large at GEO has to do with the physics of power beaming. The smaller the transmitter array, the larger is the angle of divergence of the transmitted beam. A highly divergent beam will spread out over a wide land area, and may be too weak to activate the rectenna. In order to obtain a sufficiently concentrated beam, more power must be collected and fed into a large transmitter array. Power beaming from geostationary orbit by microwaves has the added difficulty that the required "optical aperture" sizes must be very large. The 1978 NASA SPS study required a 1-km diameter transmitting antenna, and a 10 km diameter receiving rectenna, for a microwave beam at 2.45 GHz.

Chapter Six

New Technology

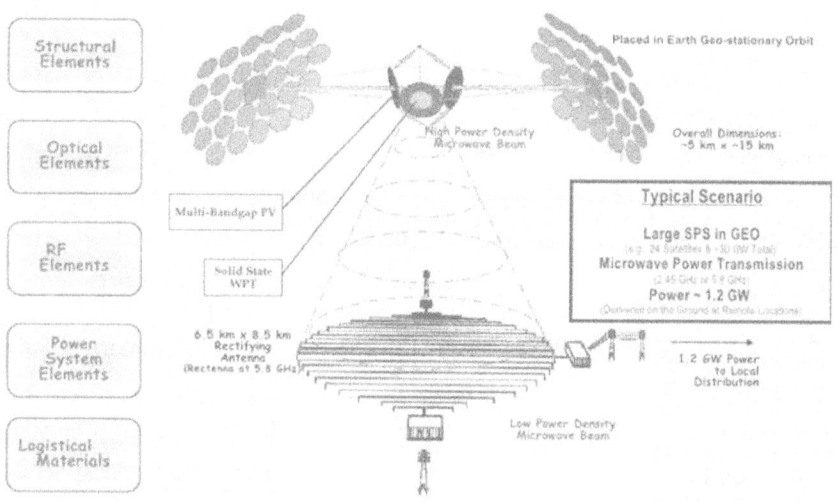

In order for SBSP concepts to become an economically viable source of clean energy, it is necessary to lower the SPS's mass and size to the point that it can be launched into working orbit with currently available or very near term commercial launch vehicles – and without requiring any on-orbit astronaut assembly. This is only possible by designing a smaller, yet more efficient, SPS system that would operate in an orbit closer to Earth. There are several ways to reduce PowerSat mass. One way would be to increase the frequency of the of microwave beam, for example to 15Ghz. This would result in some additional power losses in the atmosphere but would allow a much smaller space transmitter, thereby reducing launch costs. The transmitter mass can also be reduced by simply moving the PowerSat closer.

Another way would be to use very high levels of solar concentration. This would substantially reduce the mass of the power generating system. These ideas can all be integrated into a single PowerSat concept with a much smaller mass than any past concept.

Technologies for lower mass PowerSat:

- Higher Frequency PowerSat Transmitters.
- Transmitter Mass Reduction via Alternative orbits
- Massive Solar Concentration
- Spectrum Beam Splitting Photovoltaic Arrays

A lot of progress has been made in a wide variety of technologies and systems concepts since the last time a technology study was made on the subject back in the early 1990s. We really need to undertake a new study that incorporates these advances in technology so that we have an updated baseline model to work with. By creating an undated model we will be able to determine space solar powers economics in comparison to raising energy cost, determine the costs of developing the system and what it would cost to deploy the technology. This is one of the main reasons for writing this book, to show some of the progress that has been made and how this progress can move space solar power toward economic viability.

The biggest problem with SSP is not really technology or even economics it is the fact that the concepts that supporters of SSP are trying sale are terribly outdated. Recent papers, publications, books, magazine articles and reports coming out over the last five years are still based on PowerSat concepts that go back twenty to forty years. We need a new model based on today's technology.

Massive Solar Concentration:

Solar arrays have been and continue to be the mainstay in providing power to nearly all commercial and government spacecraft. Light from the Sun is directly converted into electrical energy using solar cells.

One way to reduce the cost of future space power systems is by minimizing the size and number of expensive solar cells by focusing the sunlight onto smaller cells using concentrator optics.

Using solar concentrator PV systems in space would potentially offer huge reductions in mass (5). At 2,000 suns concentration the solar cell mass could be reduced by some 90%. However, this mass savings would be offset by light reflector mass, which could be low mass inflatable reflectors, and radiator mass for an active cooling system.

For solar electric Space Tugs there is the possibility of using the tugs ion propellant a solar cell coolant. Since ion drives are fuel efficient the propellant would be consumed slowly, thus the propellant would generally be available for use as a cooling fluid.

As the space tug moves out into the solar system, such as a mission to Mars, the light from the sun would gradually diminish as the propellant is also diminished. Using concentrated solar on a PowerSat is a better solution as the PowerSat would remain in orbit and beam the energy to the space tug.

Looking at the NASA/DOE 1980s 5GW SSP reference models, there are actually two, the first model used silicon photovoltaic cells at 16.5 percent efficiency.

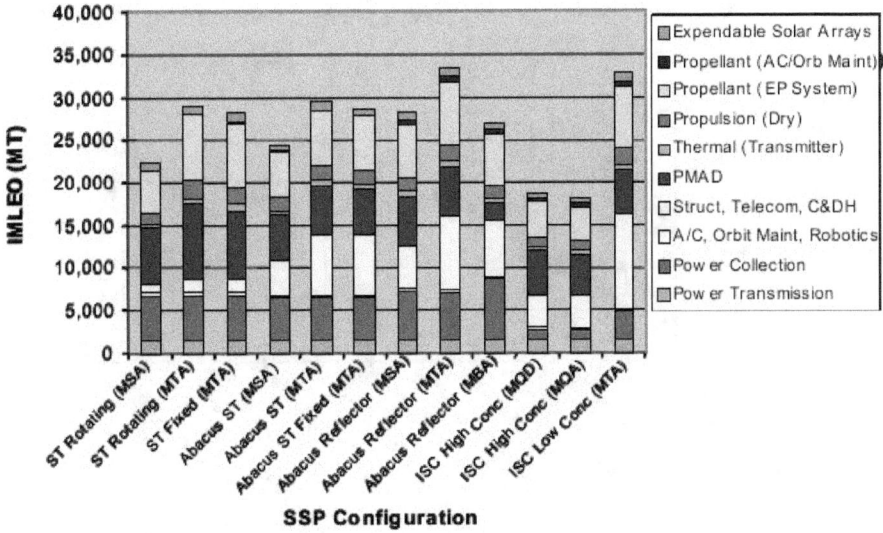

The first model had a mass of 51,000,000kg (112,200,000lbs) and the second model used higher efficiency gallium aluminum arsenide photovoltaic cells at 20 percent efficiency and 2 x solar concentrations and had a mass of 34,000,000kg (74,800,000lbs). The basic infrastructure in space or on the ground to build such massive structures in space does not exist. The cost of just the supporting infrastructure would be massive and a huge technical challenge.

The second 1980 NASA/DOE model using 2x solar concentration had only 66.6 percent the mass of the first model.

Both models provide five gigawatt of power and the transmitter is the same size on both models because the amount of power being transmitted is the same and the proposed satellite location in GEO is the same. The second model is clearly more efficient in power production and has less mass to orbit which means lower launch costs.

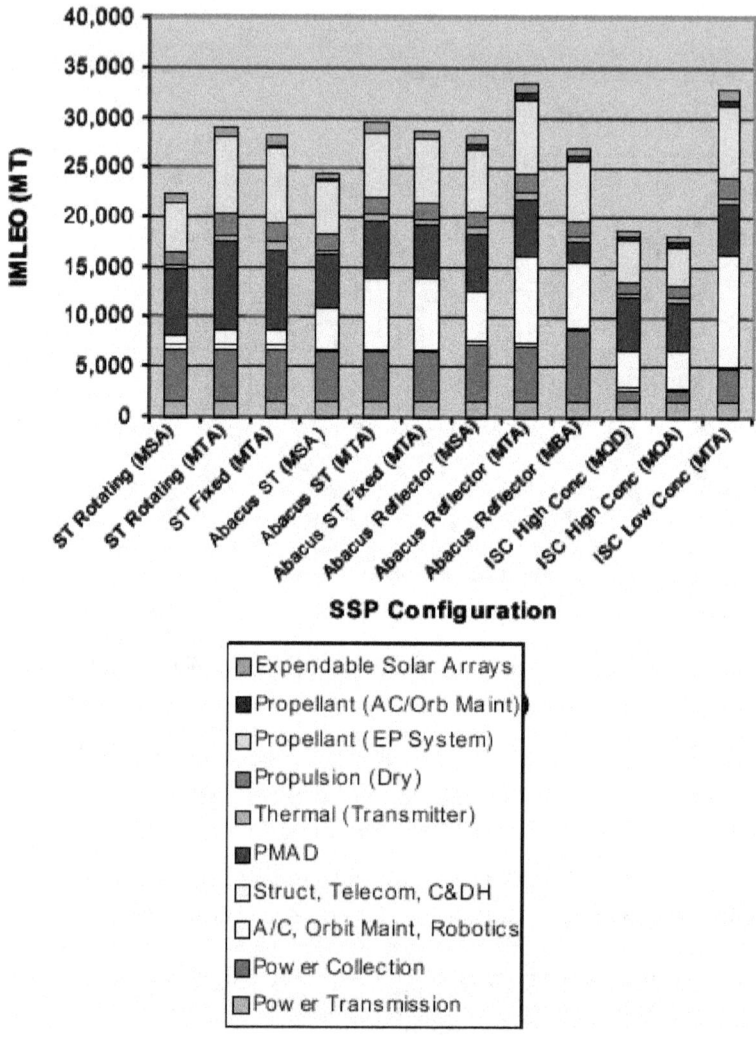

Integrated Symmetrical concentrator (ISC)

In the 1990s there were some new concepts such as the integrated symmetrical concentrator (ISC) that moved to somewhat higher frequencies (from 2.4 to 5.8 gigahertz) and higher solar concentration at 4X. This reduced the mass to 20 million kg. This was a 58.82 percent reduction is PowerSat mass. Even greater mass savings are possible if we update the technology to what is possible today.

These mass savings will allow us to consider SSP as an attractive option to replace carbon based fuels. It would allow for the first time for Earth to become an Electric Planet.

We can see that the ISC design achieved its mass reduction in large part by moving to a higher solar concentration level. We also saw this approach work well in the in 1980s models with the 2x solar concentration model. This trend is important to note because in 2008 IBM demonstrated 2300 suns solar concentration. If you are going to do SBSP you need to start by asking a few basic questions. For example, what is the trend in Photovoltaic efficiency and system design? Comparing the NASA/DOE 1980 2x solar concentration model to the flat panel model shows a substantial reduction in satellite mass (a net mass savings of 66%).

The Integrated Symmetrical Concentrator (ISC) at 4x solar concentration again substantially reduces satellite mass (another 58%). When considering the potential for concentrated solar power - What is the upper limit in solar concentration? Is it 4x, 10x, 100x, 1000x or higher? Looking at terrestrial solar power we are now seeing hardware coming to market at 500-1000x solar concentration. This technology can also be used in space. One of the great benefits of space is that there is no gravity and this allows for the deployment of very large and very low mass structures, such as large solar reflectors.

Since the ISC design concept is now very old a good place to start would be to update the model at the highest possible solar concentration levels. Once you have done that then you can look at other options such as increasing the power level, alternative orbits, deployable microwave antennas, etc., etc. The goal is to reduce the system mass to the point that it can become economically viable. How you get to that point is the challenge.

Rainbow Concentrators:

Photovoltaic arrays of the rainbow type, equipped with light-concentrator and spectral-beam-splitter optics, have been investigated in a continuing effort to develop lightweight, high-efficiency solar electric power sources. This investigation has contributed to a revival of the concept of the rainbow photovoltaic array, which originated in the 1950s but proved unrealistic at that time because the selection of solar photovoltaic cells was too limited. Advances in the art of photovoltaic cells since that time have rendered the concept more realistic. A spectral beam splitting architecture does provide an excellent basis for a multi-junction photovoltaic system with virtually ideal band gap combination, thus having the potential to reach very high conversion efficiency. In comparison to monolithically grown multi-junction solar cells, the spectrum splitting is provided by an additional optical element added into the course of the solar beam.

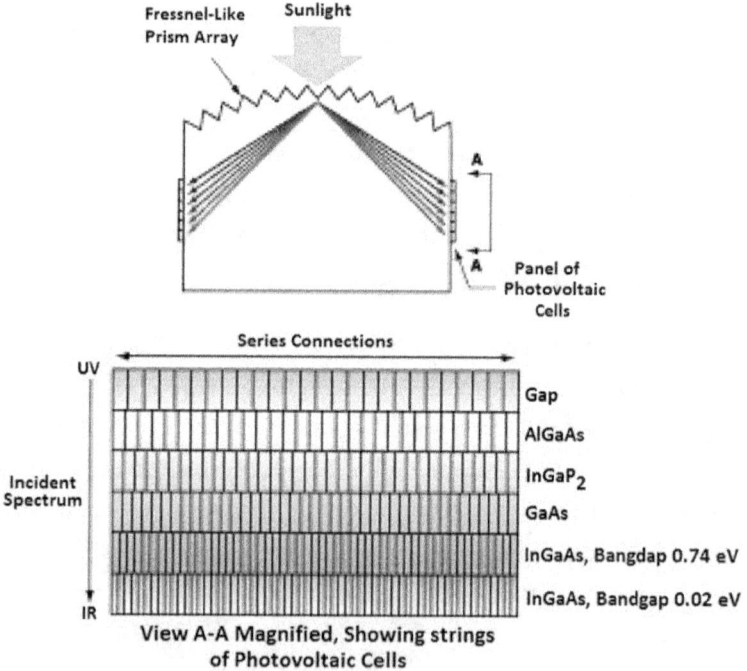

Early Rainbow Concentrator concept

The development of a "solid state" SSP system will allow deployment of high efficiency, low mass solar power satellites. To achieve this, solar radiation is first concentrated and then this concentrated beam is split into two beams using beam splitters. A beam splitter plate type is an optical window with semi-transparent mirrored coating to break a beam into two or more separate beams. One beam will contain the Infrared radiation and the other beam will contain the balance. To do this we will use an innovative system of two beam splitters, a Cold Beam Splitter and a Hot Beam Splitter. A Hot Beam splitter reflects infrared (IR) light and lets the rest pass though and the Cold Beam splitter lets the IR though and reflects the rest. This allows the use of photovoltaic solar cells that are tuned to these two groups, which can mean higher efficiency. This design was discussed in a paper written by the author and presented by Thomas Taylor at the International Space Development Conference in 2012.

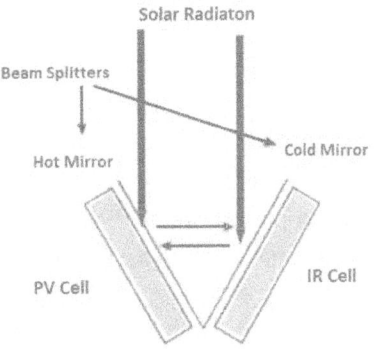

Depicts the conceptual configuration currently moving into ground testing. It consists of two different solar cells, sunlight splitting mirrors and active cooling both on the ground and in orbit. A spectral beam splitting architecture does provide an excellent basis for a multi-junction photovoltaic system with virtually ideal band gap combination, thus having the potential to reach very high conversion efficiency. In comparison to monolithically grown multi-junction solar cells, the spectrum splitting is provided by an additional optical element added into the course of the solar beam.

Using higher levels of solar concentration is very beneficial to SSP satellites due to the large potential mass reductions. The trade-off between reflector mass and solar cell mass is an easy trade as reflectors can be very low mass in space.

Using 2,000 suns solar concentration would reduce SSP satellite solar cell mass by approximately 90% verse a flat panel system with no solar concentration. This would be off-set somewhat by the increased mass of the reflectors and any required active cooling system.

In the IBM demonstration in 2008 they used a liquid metal behind the solar cell to absorb the waste heat generated by the cell. They then used water to cool the liquid metal. This technology is now finding its way into high performance personal computers as a way to cool high output microprocessors which can generate a lot of waste heat.

Solar concentration demonstrated by IBM in 2008 [14]

"IBM researchers have achieved a breakthrough in photovoltaic technology that could significantly reduce the cost of harnessing the Sun's power for electricity."

If it can overcome additional challenges to move this project from the lab to the fab, IBM believes it can significantly reduce the cost of a typical CPV based system. By using a much lower number of photovoltaic cells in a solar farm and concentrating more light onto each cell using larger lenses, IBM's system enables a significant cost advantage in terms of a lesser number of total components.

---

[14] http://www-03.ibm.com/press/us/en/pressrelease/24203.wss

IBM solar collector will concentrate the power of 2,000 suns

Modern solar collectors can concentrate only so much energy for safety's sake: too much in one place and they risk cooking themselves. An IBM-led group is working on a new collector dish that could avoid that damage while taking a big step forward in solar power efficiency. The hundreds of photovoltaic chips gathering energy at the center will be cooled by the same sort of micro channel water cooling that kept Aquasar from frying, letting each chip safely concentrate 2,000 times the solar energy it would normally face.[15]

There are advantages and disadvantages both in the use of mirrors and lenses for concentrating sunlight. Lenses are associated with a relatively high optical loss, typically about 30%, due to reflection and absorption. The heat load is not a problem in lens systems because the cells are spread out on a plate, with plenty of space for each cell to dissipate the waste heat. With mirrors the typical losses are only 10-15%. However, because mirrors focus all of the light onto one highly illuminated area, the cells must be placed closely together in an array.

---

[15] http://www.engadget.com/2013/04/22/ibm-alliance-solar-collector-concentrates-power-of-2k-suns/

All of the heat load must therefore be dissipated through the back of the cells, which makes cooling in these systems a challenge. In fact, the efficient removal of this heat load is one of the major obstacles for creating viable mirror-based high concentration systems. There is a need for a cooling device that uses a liquid coolant, cools efficiently across the entire surface, while not shading the concentrator, and operates at a low pumping power.

There is no particular reason why you can't use this or similar technology with an SSP satellite. The main difference would be that in space you would probably use a different coolant fluid to cool the cells. This coolant could be ammonia for example. Another possibility would be to use the electric propulsion system propellant, such as Xenon or Argon, as the coolant. This would allow dual use of the propellant.

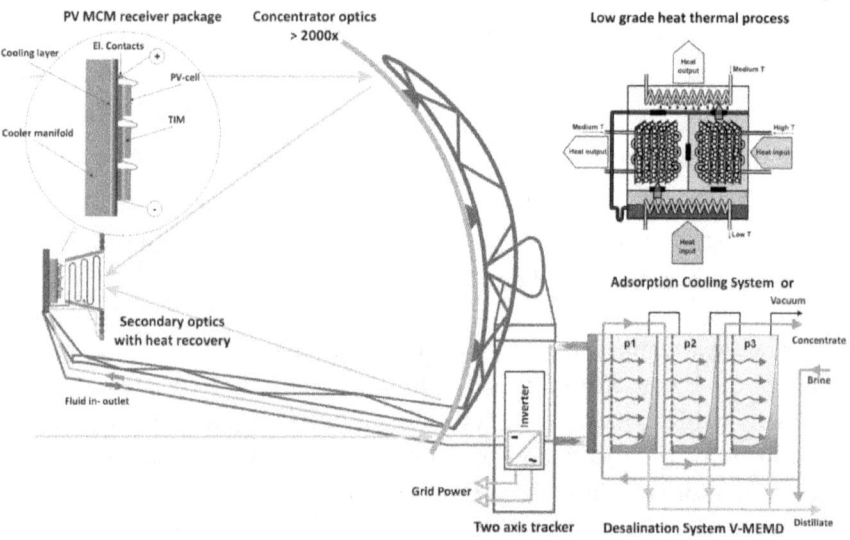

This image shows a solar concentrator using water cooling. In space you might use ammonia as the coolant.[16]

---

[16] C. Lee, W. Escher, S. Paredes, (2012), A novel concept of energy reuse from high concentration photovoltaic thermal (HCPVT) system for desalination, Elsevier Science, Volume 295, Pages 70-81

Since the propellant is consumed slowly over a period of years this might work out really well. In space it is easier to get rid of high temperature heat than low temperature heat. An effective means for transferring thermal energy is with a pumped fluid loop. A pump provides the driving force necessary to create fluid flow over a hot surface of the solar cells transferring heat to the fluid and subsequently over another, colder surface that accepts heat from the fluid. A variable speed pump can be included to throttle flow based on temperature requirements of the system.

In space it can get terribly cold, especially when shadowed from the sun's rays. Even so, it can be difficult to get rid of heat in space since there is no air or other mechanism to promote convection of the waste heat away from the cells. Low mass passive and active cooling system solutions will be needed. One possible passive solution would be to get rid of the Ultraviolet (UV) rays before they reach the solar cells. This might be done by absorbing the UV radiation at the large light reflectors and then radiating it out into space. UV is only 3% of the solar spectrum but is almost 100% waste heat at the solar cells, so getting rid of it before it strikes the cells can reduce the amount of waste heat.

Photovoltaic Mass Reduction:

In 2008 IBM demonstrated concentrated solar at 2,300 suns, receiving 230 watts/sq/cm solar input an generating 70 watts electrical output for an efficiency of 30%. Using a cascading PV system it would be possible to obtain an efficiency of 40% or 92 watts/sq/cm electrical output. At 92 watts/sq/cm we could obtain 92,000 watts per square meter. 100 square meters of cells would generate 92,000,000 watts or 92 megawatts of electricity. Using advanced triple-junction cells at 84mg/cm the mass of the 100 sq/m of PV cells would be approximately 84kg.

Solar concentrators for use in space have received growing attention in the past few years in view of their many potential applications. Among those, perhaps the most important ones are space power generation and solar thermal propulsion.

In the former, the concentrator is used to focus solar radiation on a conversion device, e.g., a photovoltaic array or the high temperature end of a dynamic engine. For GEO you will notice that an SSP system using 40% efficient solar cells at 2,000 suns concentration might reduce satellite mass by some 70% and reduce LEO to GEO ion propellant mass by 552,924kg (70%). Note that reflector technology for 10,000:1 concentration has already been proposed for Solar Thermal Rockets by both NASA and DOD.

What is the current state of technology in concentrated solar power? What if the mass of SBSP power system could be substantially reduced using massive solar concentration? If the SBSP Satellite uses solar concentration at 2,300 suns we can also reduce the power system components by an order of magnitude or 90% (IBM, 2008), but this would be offset somewhat be the addition of reflector mass and the mass of an active cooling system.

Results of the IBM solar experiment at 2,000 suns concentration

2,300 suns

230 watts onto a centimeter square solar cell

Cell size : 1cm/2

Cell mass : 1mg/cm

70W of usable electrical power

30.4% efficiency

Waste heat : 69.6%, 160.08 watts

Un-cooled temp 1600°C

Cooled temp 85 °C

Cooling technology: liquid metal/water

Let's that a look at how this would affect PowerSat mass. The reduction is solar cell and supporting structural mass could substantially reduce the Mass of the PowerSat.

5.8Ghz 2,300x Concentration, 30.4% efficient PV, 1GW, Mass 7,640,000kg,

Transmitter 6,500,000kg, Power 1,140,000kg

Note that the transmitter is now 6 times more massive than the power system.

SBSP Application 2.5GW:

2,500,000,000 watts / 70 watts per cell = 35,714,285 cells

35,714,285 cells x 1mg each = 35,714,285 mg = 34.714 kg

A 2.5 GW PowerSat operating at 2,300 suns concentration would only need 34.7 kg of solar cells.

Photovoltaic Cooling:

For cooling the photovoltaic cells a wide variety of systems concepts are possible. One particular setup proposes to use an active cooling system using Ammonia similar to the one used on the International Space Station (ISS). Ammonia freezes at -107 degrees F (-77 C) at standard atmospheric pressure. The heated ammonia circulates through huge radiators located on the exterior of the Space Station, releasing the heat as infrared radiation and cooling as it flows.

The heat-bearing ammonia can't lose heat fast enough to reach its freezing point before the liquid circulates back inside the warmer confines of the satellite. The Ammonia will pass between the sets of solar panels and extract the waste heat. The Ammonia is then cooled and recycled back to collect more waste heat. The proposed concept works well for space solar satellites as well.

The Holy Grail of SSP:

Concentrated Solar Power (CSP) is the Holy Grail of Space Solar Power (SSP). The key to accessing the Holy Grail is Thermal Waste Power Management. When considering for example the Integrated Symmetrical Concentrator (ISC) SSP systems concept as presented the presentation of on H. Feingold, C. Carrington, Evaluation and Comparison of Space Solar Power Concepts, Presented at the 53rd *International Astronautical Congress,* 2002,IL,USA17

We see two models presented. The first model uses a solar concentration at 2x and the second model uses solar concentration at 4x. When compared to the NASA/DOE 1980 2x model we see that the benefits of the 2x ISC model is just an improvement in packaging as the separation of the solar concentrator from the PV array allows the PV cells to be packaged much closer which reduces wiring and structure. The 4x ISC is however a large improvement over the 1980s model (and the ISC 2x model) because the higher concentration ratio actually reduces PV mass and the wiring and structure that goes with that mass.

---

[17]http://www.spacefuture.com/archive/evaluation_and_comparison_of_space_solar_power_concepts.shtml

The logical question to ask at this point is - What would happen at high solar concentration levels? The answer is of course greater reductions in PV, wiring and structural mass. At 2,000 suns concentration there is an order of magnitude reduction in solar cell mass. The problem that comes into play is that high solar concentration ratios increase the Thermal Waste Power Management problem. Additional mass has to be added to the CSP system to handle this problem. Designing a low mass thermal solution to space CSP is the key to the Holy Grail.

Carbon/Carbon radiator:

The carbon-carbon radiator panel conducts thermal energy more efficiently than other materials currently being used to dissipate thermal energy on satellites. The use of high conductivity fibers in C-C fabrication yields materials that have high stiffness and high thermal conductivity, and since the density of C-C panels is considerably lower than aluminum panels typically used on satellites, significant weight savings can be realized by replacing Aluminum radiators with C-C radiators. C-C also has an advantage over other high conductivity composites materials in that the thermal conductivity through the thickness of the material is significantly higher. C-C also has markedly higher specific thermal efficiency than aluminum and offers improved performance for lower volume and mass.

The advantages of carbon-carbon composite are:

- it has high strength
- it has low weight (density less than aluminum)
- it has high thermal conductivity
- it can be used at high temperatures (currently used in aircraft brakes and the Space Shuttle wing leading edge).

Another possible solution to waste heat management would be to use the empty rocket stage propellant tanks as space radiators.

SLS Configuration

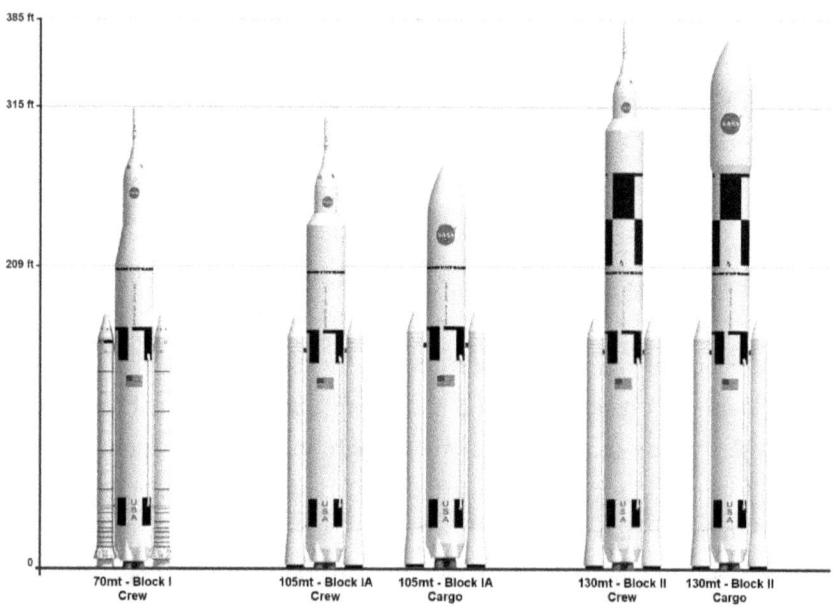

Space Launch System (SL): Note the huge size of the vehicles tanks. SLS can lift approximately 80,000 kilograms into orbit.

Once launched into space the chemical propellant tanks would be vented and then used as radiators. Since the mass would be essentially free on orbit this appears to be a good approach to reducing radiator mass.

Since heat rejection at high temperature works better in space you might want to run the coolant through the space radiators first and then dump it into the launch vehicle tanks and use them as medium to low temperature radiators. Here the coolant would come into contact with the tank walls and radiate the heat out into space. For the space radiators a Carbon/Carbon radiator could be used.

NASA's Space Launch System could deliver multi-megawatt PowerSats into low Earth Orbit. Such PowerSats could beam energy to several space tugs traveling back and forth between the Earth and the Moon. This would provide a low cost in-space transportation to support Lunar or free space colony settlement.

Such a system could also be used as a booster system for crew and cargo transport to Mars. Increasing energy consumption, shrinking resources and rising energy costs will have significant impact on our standard of living for future generations. In this situation, the development of alternative, cost effective sources of energy has to be a priority.

It is possible to design, engineer and deploy the proposed PowerSats within just a few years. The basic technology already exists and this technology when combined with the innovation offered by the authors makes SSP an economically attractive technology for near future energy production. In the future new heavy lift launch vehicles, such as SLS could deploy ever larger versions of these PowerSats, possibly of gigawatt scale as technology reduces the mass of the PowerSats.

Chapter Seven

Alternative Orbits

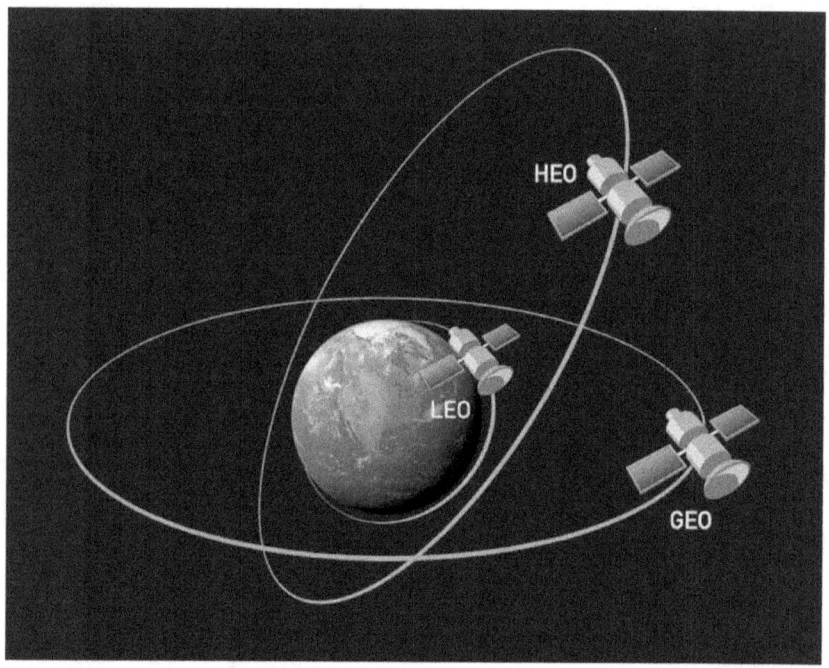

Accelerating the development of space based solar power is important to the future of mankind. A continuous and clean source of energy is greatly needed to sustain growth and for the protection of the Earth's environment. Innovative approaches to in-space power production have the potential of offering cost effective supplies of power.

The current baseline concepts specifying GEO PowerSats are flawed. They are poor design concepts that do nothing to advance SBSP development – because they put forth uneconomical and unaffordable system concepts. These concepts actually discourage SBSP development.

Until we develop new, more realistic and more economically viable systems concepts for solar power satellite implementation we will be wasting our time trying to sell it to governments and to the people of this planet. By establishing economically viable models for SBSP we can move closer to pure green energy and accelerate man's move into space at the same time.

The highest priority research areas for solar power satellites are those where major improvement can be made in the technical feasibility and/or cost of the system. The advantages of space-based solar power cannot be realized in the near-term due to the presumed cost of transmitting power from orbit to receiving stations on Earth. These two components are interdependent due to the need for high efficiency power transmission. Since SBSP microwave transmitter size and mass is a direct function of distance between transmitter and receiver, only sub-GEO satellites should be considered. This can be a shocking revelation for people who have always taken it as a given that SBSP satellites must be positioned in a geosynchronous orbit.

For the last three years I have been focused on developing alternative solutions for Space-based Solar Power (SBSP). This started with an extensive review of past concepts to determine why SBSP remained uneconomical forty years after its invention by Dr. Peter Glaser. The conclusion was that locating the SBSP Satellites in GEO posed a substantial mass penalty on the Satellites due to the need for a very large satellite transmitter and the need for in-space transportation from low Earth orbit (LEO) to geostationary orbit (GEO).

So the question then became – If not GEO where?
Geostationary satellite communication systems have clear and undeniable cost and simplicity advantages. With thousands of ground antennas in satellite television systems, the stationary ground antennas become a massive cost advantage over non-geo systems. However, the AIAA has recently reported that over 1000 satellites exist in geostationary orbit. More than 700 of these are uncontrolled.

The environment may become physically crowded within the coming decade. The environment also may soon face electromagnetic crowding problems, with typical spacing at 2 degrees of orbital arc.

So It has been indicated that a careful choice of Molniya eccentricity near 0.722 would allow the Complex, expensive AZ-EL ground antennas to be replaced with simple relatively inexpensive Single-axis Brandon antennas. The Brandon antennas would be expected to offer attractively High data rate, in the neighborhood of that required for satellite-ground TV[18].

When considering options for SSP there several alternative orbital locations could be considered. These include:

- Low Earth Orbit
- Molniya
    - Sub-Molniya
- Medium Earth Orbits
    - Sweet Spot (6-hour orbit) Geostationary
    - 5,100km SS-O
- Space Grid (SS-O with Equatorial microwave reflectors)
- Geostationary

The orbital location of the SBSP Satellite(s) will determine the mass of the SBSP satellite transmitter. While most past concepts placed the PowerSat in GEO there are other perhaps better locations. One possible location would be a 6-hour orbit at 10,300km. This orbit lies between the inner and outer Van Allen radiation belts (the sweet spot) and is only 28.6% of the distance of GEO.

---

[18] Paul Christopher, Molniya System Alternatives for Geostationary Satellite Systems, VA, USA

This orbit would have approximately 12% shadow time per orbit. Another alternative would be the Space Grid. The Space Grid uses a sun-synchronous near polar orbit at 800km and therefore never enters the Earth's shadow. The energy produced by the PowerSat is beamed to microwave reflectors in equatorial orbit and reflected to the rectenna on the Earth's surface. Another version of the Space Grid places the PowerSat in a 2,722km sun synchronous orbit at 110 degrees that is closer to the reflectors and therefore has a shorter beam distance. This allows for an even smaller transmitter.

Potential PowerSat Locations:

Location	Sat Distance	Beam Distance	Factor	Squared	TxMass
GEO	36,000	36,000	1	0	13,000,000
Molinya 12-hour	12,000	12,000	3	9	1,444,444
Molinya 3-hour	8,000	8,000	4.3	18.49	64,1975
MEO Sweet Spot	10,300	10,300	3.49	12.18	1,067,323
MEO SS-O	5,100	6,000	7.05	49.76	261,254
LEO	1,100	1,000	32.72	1070.83	12,140
Space Grid	2,722	8,000	4.5	20.25	641,985

As shown the inverse square law works in favor of closer SBSP satellites.

As we can see from the chart above the Sweet Sport has low shadow time and the Space Grid concepts have no shadow time at all. So, the ideal that PowerSats have to be located in GEO to obtain low shadow is simply not true.

Transmitter mass is a large part of SBSP satellite design and therefore affects the system cost, especially launch costs.

Since the power transmission subsystem represents about half the capital cost of the total SPS reference system, it is worthwhile to consider the lower orbit alternatives so the technological, environmental, social and political problems and relative advantages may be assessed in comparison with those of geosynchronous forms. Thirty six thousand kilometers above earth would seem to be a logical destination for a number of reasons, but that orbit is already largely committed. What is more, this great height and the mass and number of space solar systems proposed for GEO will not be cost-justifiable anytime soon. Decades will pass before this promising location will be a major solar power satellite (SPS) destination due to incumbent player resistance over possible signal interference. Also, dramatic improvements in space-based PV cell technology will be needed, as will reductions in the cost of space launch.

SPS systems will be a predictable contributor to our energy future when PowerSats are built to operate in space at costs competitive with energy systems on Earth. Successful SPS designs will be those that are technically feasible, economically affordable and can be proven to work.

Toward this end, the authors propose the development of a SBSP system using lower mass solar power satellites working from closer orbit. The closer proximity of the satellite transmitter and ground receivers allows the satellite transmitter size and mass to be much lower. A SBSP system utilizing a closer orbit can be constructed on Earth and launched via available launch vehicles - with no on-orbit astronaut assembly required. This will remove a huge cost burden from the system.

The argument against LEO and MEO PowerSats has mostly to do with power losses due to Earth shadowing which would require larger satellites or more satellites to make up for the lost shadow time.

The assumption has been that this would result in a total mass greater than that of a GEO PowerSat. Is this argument valid? Actually, no it is not. While it would add mass in some cases, the mass would still be lower than that of a GEO PowerSat in most all cases. Therefore, the argument is false. Now let's prove it.

Using the 1980 NASA/DOE 16% efficient 5 GW flat panel model and assuming the same energy level at 5GW for the power systems for all models at 38 million kg we have.

Location	Distance	Transmitter	Power System	Total
GEO	36,000	13,000,000	38,000,000	51,000,000
Molinya 12-hour	12,000	1,444,444	38,000,000	39,444,444
Molinya 3-hour	8,000	641,975	38,000,000	38,641,975
MEO Sweet Spot	10,300	1,064,174	38,000,000	39,064,174
MEO SS-O	5,100	260,902	38,000,000	38,260,903
LEO	1,100	12,137	38,000,000	38,012,137
Space Grid	8,000	641,975	38,000,000	38,641,975

With transmitter mass based on distance we can see that the transmitter mass in all scenarios closer than GEO is less. Propulsion mass is estimated at .05% of PowerSat mass. However, we have to add mass to offset shadowing in some cases.

Location	%Shadow	Added Mass	Sub-total	Propellant	Total
GEO	0	0	51000000	2550000	53550000
Molinya	33	13016666	39444444	2623055	55084167
Sub-Molinya	33	12751851	38641975	2569691	53963519
Sweet Spot	12	4687700	39064174	2187593	45939469
MEO SS-O	0	0	38260902	1913045	40173948
LEO	40	15204854	38012137	2660849	55877842
Space Grid	0	0	38641975	1932098	40574074

Some concepts such as the Space Grid don't require additional mass for shadowing but require additional mass for power distribution using reflectors. Reflector mass is estimated at .05% of PowerSat mass. The MEO Grid requires both.

Location	Mass	Reflector Mass	Total
SS-O Grid	40,574,074	2,028,703	42,602,777

The idea that GEO is the best location based on mass to orbit has been debunked. Now let's go a little further. The problem with recent studies is that they make little effort to update the state of technology for SSP. So, let's do a minor upgrade and simply increase the solar cell efficiency of the 1980 NASA/DOE flat panel model from 16% to 32% (no solar concentration – yet). This would reduce the power system mass from 38 million kg to 19 million kg. How would that affect these scenarios?

Location	Transmitter Mass	Power System	Total
GEO	13,000,000	19,000,000	32,000,000
Molinya	1,444,444	19,000,000	20,444,444
Sub-Molinya	641,975	19,000,000	19,641,975
Sweet Spot	1,064,174	19,000,000	20,064,174
MEO SS-O	260,902	19,000,000	19,260,903
LEO	12,137	19,000,000	19,012,137
Space Grid	641,975	19,000,000	19,641,975

Even after adding in additional mass for shadowing or SPR reflectors, and in the case of the LEO/MEO Grid both, all scenarios have less mass than GEO. Lowering the mass of the power system has a larger effect on the non-GEO scenarios because in-space transportation propellant is substantially lower. Now let's really up-date the scenarios and see what happens at 2000 suns solar concentration, including transfer propellant and add in for shadowing and reflectors as needed.

Location	Transmitter	Power System	Total
GEO	13,000,000	1,900,000	14900000
Molinya	1,444,444	1,900,000	3344444
Sub-Molinya	641,975	1,900,000	2541975
MEO (Sweet Spot)	1,064,174	1,900,000	2964174
MEO SS-O	260,903	1,900,000	2160902
LEO	12,137	1,900,000	1912137
Space Grid	641,975	1,900,000	2541975

The effect of reducing power system mass by using massive solar concentration has improved all the scenarios, but notice that one of the worst scenarios before is now one of the best – LEO. The transmitter mass reductions combined with power system mass reductions result in huge mass savings for close in systems, so huge that they become one of the best options for SSP – even after taking shadowing into consideration. To be fair there are also other issues such as the oceans and beam time to target which also effect LEO location.

There is also the issue of the minimum size of a PowerSat and this is also related to beam distance.

Location	Minimum Size	#Sats	Mass per Sat
GEO	15,645,000.00	1	15,645,000.00
Molinya	4,670,516.67	9	518,946.30
Sub-Molinya	3,549,868.52	20.25	175,302.15
MEO (Sweet Spot)	3,485,869.07	12.216	285,351.74
MEO SS-O	2,268,947.92	49.827	45,536.52
LEO	2,810,841.90	1071.1	2,624.32
SS-O/Equatorial Grid	4,799,212.96	20.25	236,998.17

It can also have a huge impact on launch costs. Below we assume a near term launch vehicle with 50,000kg lift capability and a cost of $100 million per launch.

Location	Mass	Payload kg	Launches	Launch Cost	Total
GEO	15,645,000	50000	312.90	100000000	$31,290,000,000
Molinya	4,670,517	50000	93.41	100000000	$ 9,341,033,333
Sub-Molinya	3,549,869	50000	71.00	100000000	$ 7,099,737,037
Sweet Spot	3,485,869	50000	69.72	100000000	$ 6,971,738,148
MEO SS-O	2,268,948	50000	45.38	100000000	$ 4,537,895,833
LEO	2,810,842	50000	56.22	100000000	$ 5,621,683,796
Space Grid	4,799,213	50000	95.98	100000000	$ 9,598,425,926

It appears that you could pay for the entire MEO SS-O constellation just from the launch cost savings verse GEO.

GEO:

Most past concepts relied on putting the PowerSat is GEO because it could provide constant power and would stay positioned over the receiver on the ground. However, SSP is simply not economically feasible using the GEO/PV systems that the space community keeps trying to sale to the government and general public.

This is a promising avenue for future work, but the performance, mass and cost of these proposed systems do not yet permit a cost-effective system. There is no research that suggests that it is cost effective or will be cost effective for perhaps decades. For GEO/PV to be cost effective you need a massive improvement of PV cell technology and a massive decrease in the cost of space launch.

While possible, these are not likely to happen within a generation. It is time to give up the unaffordable simple elegance of GEO and begin the hard work on viable Space Solar Power system concepts. The future of SSP satellite systems will be related to their cost-efficiency. The probability of SSP becoming a part of our energy future will increase if the system can provide costs competitive with other energy systems. They must be economically feasible and economically affordable. This is only possible by designing a smaller, yet more efficient, SPS system that would operate in an orbit closer to Earth.

The past assumption was that GEO was the best location was based on 1) availability of power and 2) the ability to beam to a single location 24/7. As I have already shown, there is at least on other location other than GEO that can provide power 24/7 power and that there are at least two ways to use this location. This location is a sun-synchronous dusk to dawn orbit.

This obit can be 800km or 1,500km and is therefore very close to the Earth. This orbit can be used with equatorial or Molniya based reflectors to deliver 24/7 power. But let's really dig into the argument and take a close look at how distance affects the transmitter mass.

There is no logic to placing SBSP Satellites in GEO. The only advantage is the ability to constantly remain over one area on the Earth's surface; however, this advantage is not worth the massive increase in satellite mass that this single advantage gives. For example, the transmitter mass difference between GEO 24-hour orbit and a circular 12-hour orbit is 50%.

The mass of the transmitter is cut in half simply by moving it closer to Earth, the ground receiver is half the size and the system is more economically viable.

The mass requirements for GEO SBSP make such a system economically unviable. That is why every study of GEO SBSP qualifies its viable based on some futuristic launch vehicle that does not exist and may not exist for a very long time. Therefore, if SBSP is going to happen you have to consider options other than GEO. This might be a sub-Molniya orbit (3 or 4 hour orbit) or an equatorial circular orbit, or something else, but it will not be GEO.

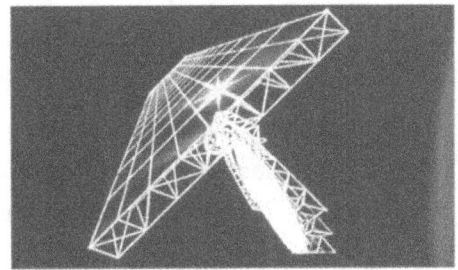

Orbits: SPS in Geosynchronous orbit. ~1980 Reference System

The only real benefit of GEO is the ability to stay over the target receiver constantly. The major negative points of GEO are 1) Transmitter size – this is a major issue and possibly the killer for GEO. 2) LEO to GEO transfer – the cost of this transportation is very substantial, past concepts including ion propulsion, which is the best approach still have huge costs in propellant mass to orbit. 3) GEO is already highly populated with ComSats – even if you could solve the other problems this would still kill any GEO SBSP satellite proposal.

"A detailed analysis may very well show that building one large broadcast antenna for a GEO sat would be cheaper, mass wise, than building 2 small broadcast antenna for a LEO sat in order to make up for the slew and acquisition requirements. "

Given the mass requirement for a microwave transmitter in GEO this is highly unlikely. Using two transmitters could actually be beneficial by allowing service to more ground receivers. A 100mw satellite could be beaming 50mw to one receiver and 50mw to another receiver at the same time.

To overcome these problems we use Sun-synchronous orbit (SS-O) PowerSats and a space power relay (SPR). These concepts will be discussed briefly and then merged into a new concept call the Space Power Grid.

Using a powerful focused beam in the microwave range, long distances can be covered. There are two methods of wireless power transmission for bridging applications. First is the direct method, from transmitting array to rectenna. The second method is via a relay reflector between the transmitter and rectenna. This reflector needs to be at an altitude that is visible for both transmitter and rectenna. A plane or elliptical reflector can be used, but needs to be at an altitude were the reflectors are able to see the rectenna.

Low Earth Orbit:

The largest potential application for microwave power transmission is SBPS satellites. In this application, solar power is captured in space and converted into electricity and beamed to the Earth. Several concepts have been proposed in the past for LEO PowerSat beaming to Earth to alleviate the launch cost problem. It has been known since at least 1980 that placing PowerSats in LEO would reduce satellite transmitter mass by "an order of magnitude" (Drummond (2))), i.e., about 90%. However, the problems of PowerSat stationing in LEO are the Earth's rotation under the satellite, Earth shadowing and the Oceans.

Concepts have been proposed for low-earth-orbit (LEO) satellites beaming power to earth, to alleviate the launch cost problem. This raises the difficulty of short, intermittent transmission and active tracking; however technological advances have made these acceptable. The shorter transmission distance (1,100km to LEO vs. 36,000 km to GEO) greatly reduces antenna size and mass.

Soubel (2004) and JAXA(2004) describe a Japanese project to construct a 50KW demonstration satellite in LEO followed by a 10MW satellite in a 1,100km-high orbit, giving 200 seconds of power to ground receivers during each pass, with an orbit period of roughly 90 minutes, with retro-directive power beaming to enable tracking. This is proposed for retail power beaming to devices such as cellphones. Hoffert (2004) proposed an evolutionary technical approach to space solar power demonstration, stepping up through terrestrial

point-to-point beaming, intermittent beaming from LEO to consumers in developing nations combined with storage devices, and beaming/reception demonstrations to high orbits using large facilities such as the Arrecibo radio telescope. The evolution here was however limited to government-funded confidence-building for full-scale SSP rather than early revenue, so that the basic SSP problem of the huge cost-to-first-power remained. [19]

The benefits of SPS deployments in LEO impact not just the space segment, i.e. the space transmitter, but also the ground receiver. According to Kotin, writing in 1978, the total land area required by each rectenna facility, including provision for a microwave buffer zone, based on GEO-located satellites is estimated at approximately 50,000 acres or 200 square kilometers. (Kotin, 1978) By locating the satellites in LEO ground receiver size is reduced by over 90 percent. Past cost estimates for ground systems using the GEO satellite reference exceed $2 billion. Alternately, LEO satellite system ground receivers using the Sunflower concept will require more or less 4 square kilometers of space, costing a small fraction of the GEO system ground receiver.

---

[19] An Evolutionary Model for Space Solar Power, Nicholas Boechler, Sameer Hameer, Sam Wanis, Narayanan Komerath, School of Aerospace Engineering, Georgia Institute of Technology, Atlanta GA 30332.

The SBSP LEO PowerSat design is such that is allows power to be produced while in sunlight which is approximately 60 percent of each orbit. The solar radiation is collected via a reflector(s) and photovoltaic (PV) array(s). The envisioned array is 2-sided allowing for two reflectors rather than one large one. Additionally, making the array 2-sided allows for easier cooling of the PV array. Energy produce is stored on-board the satellite using Flywheels.

Atlas V launch Vehicle

Flywheels where chosen because while batteries can provide the energy density they cannot provide the large number of cycles required by this design. Capacitors can provide the cycles but not the energy density required. Flywheels can provide both the energy density and the cycles needed for an equatorial low Earth orbit (ELEO) system.

Another problem with past SBSP Satellite designs was that they were huge and required new, large and super cheap rockets to put them into space (These do not exist). To make the satellites more affordable I broke the large mega-satellite it into smaller independent, free flying satellites of only 20,000kg each. These could then launch on existing launch vehicles such as the Delta IV, Atlas V, Ariane 5 and Proton.

Future heavy lift launch vehicles can be utilized when they become available. After doing all of this I discovered that while this was a tremendous improvement the satellites where still not profitable. This should give you an idea of just how hard developing SBSP really is and why it has never been developed. The skeptics have always been right. I needed more innovation. I had noticed a trend in increased solar concentration in past designs. In the 1970s there was a proposal for 2x solar concentration, which reduced the original NASA base-line design by about thirty one percent.

Then in the 1980s there where proposed designs at 4x solar. Looking at this trend I decided to research the state-of-art in solar concentration and found that in 2008 IBM has demonstrated 2,300 suns concentration using personal computer (PC) cooling technology to keep the photovoltaic cells from overheating. I then applied a 2,000 sun solar concentration level to the Molniya SBSP Satellite design and found that it was not only economically viable but potentially highly profitable. We can also apply this technology to an LEO constellation. A quick and dirty look at potential orbital locations suggests that LEO based SSP systems combined with high solar concentration levels would be the lowest mass option for SSP. This appears to be true even after adding in additional mass to offset the Earth's shadowing of the satellite.

Orbit	Distance km	Solar	PV Mass	Freq	Tx Mass	Total Mass
GEO	36,000	1	12120000	2.4	13000000	25,120,000
GEO	36,000	2,000	1212000	5.8	6500000	7,712,000
LEO	1,100	2,000	1,212,000	5.8	92857	1,304,857

Considering the number above we see that LEO has the lowest mass. However, we have to add more mass to make up for the 40% loss due to Earth shadowing.

To overcome the lost power generation caused by Earth shadowing more or larger satellites have to be placed into orbit. For LEO satellites this can be a much as 40% more mass.

> 521,943 kilograms

Even after adding in additional mass to off-set Earth shadowing we see that the total mass is less than GEO, and this does not even include mass to orbit savings due to less in-space propellant use.

There is another interesting aspect in locating the PowerSats in LEO and this is the minimum beam density. Since the PowerSat is much closer the minimum size of the PowerSat can be substantially reduced. By locating the PowerSat in a 1,100 kilometer orbit we can deploy PowerSats as small as 200 megawatts. Let's take a look at this.

5000000000 / 200000000 = 25

We could deploy 25 smaller PowerSats and receive the same net energy and each PowerSat would have a mass on only 73,072 kg.

We would still have some problems with the constellation, including, limited beam time due to the speed of the PowerSat and the large size of the Oceans.

Due to the mass-to-orbit requirements for geosynchronous SPS, other options should be considered. Moving the satellite closer allows for much small transmitter arrays, but this causes two new problems, Earth shadowing and reduced beam time due to the speed of the satellite. The beam time would only be about 300 seconds per orbit per rectenna. With an Orbit Revolution Time of 107 minutes, it would require 107 x 60 = 6420 sec / 300 sec = 21.4 rectenna spaced equally around the Earth. This might be possible but would be difficult given the size of the Earth's oceans. There are some possible solutions such as using the energy for LOX/Hydrogen or distilled water production, or simply skip a couple of areas. Another possible option would be to use the onboard power storage. If the power storage used flywheels and the power storage density where increased to 400 or even 500w/kg this would allow greater spacing between the ground sites. Considering the large mass reduction possible with LEO PowerSats we could consider on-orbit power stored possibly using flywheels to reduce the number of ground rectenna.

While it is possible to consider thermal or electrical power storage, the problem of Earth Shadowing will not be address and will be treated as part of the system design, i.e., no power collection during that part of the orbit.

Let's take at this concept is some detail and see how it works out.

Orbit: 1,100 km

6426.62 seconds / 3600 seconds (1 hour) = 1.7 hours

Power per Satellite: 280 Mw

Orbit Time: 6426.62 seconds

Beam Time: 200 seconds

Number of Rectenna: 3 (Because of the large size of the Oceans we will only use three sites).

Total beam time per orbit 3 x 200 = 600 seconds

Non-beam time per orbit: 5,626.62

Number of Satellites in Constellation: 25

We can see that each satellite can dump its energy three times on each orbit. Therefore, we would need each satellite to store energy using the Flywheels until it passes over a ground receiver. The primary problem with LEO PowerSats is beam time. The beam time is so limited that it is difficult to get the power to the ground. Power storage is to massive with today state of technology.

Molniya:

One way to shorten time-to-term, and thereby alleviate some of these constraints, will be to look for a workable non-GEO orbit. The author (Jones, R.) originally suggested a highly elliptic 3-hour orbit similar to the Molniya orbits used by the Soviets only operating much closer to Earth.

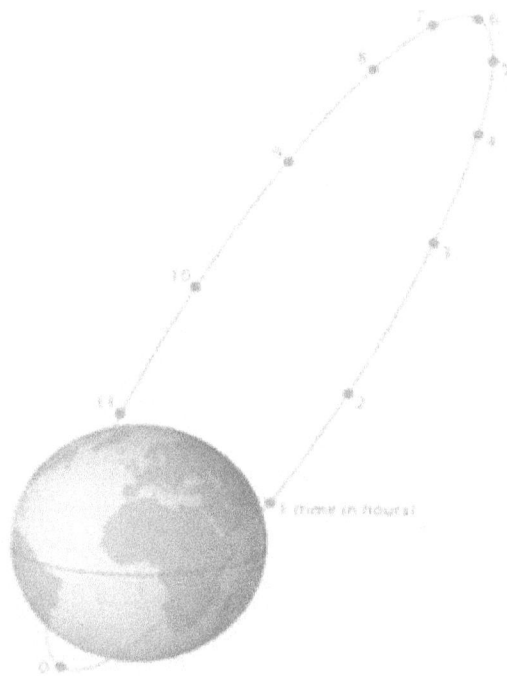

This orbit will provide about two hours of transmission time per orbit and one hour of non-transmission time. When compared to the GEO location, this new class of sub-Molniya orbits has the potential to substantially reduce SSP satellite mass.

Molniya satellites, with inexpensive ground antennas might be an attractive Alternative to predominantly GEO systems with declining Real Estate. There also other options, such as a 6-hour equatorial orbit and the Space Grid.

The size of a solar power satellite would be dominated by two factors: the size of the collecting apparatus (eg. panels and mirrors), and the size of the transmitting antenna. The distance from Earth to geostationary orbit (22,300 miles, 35,700 km), the chosen wavelength of the microwaves, and certain laws of physics (specifically the Rayleigh Criterion or diffraction limit) will all be factors.

Power transmission of tens of kilowatts has been well proven by existing tests at Goldstone in California (1975) and Grand Bassin on Reunion Island (1997). At the Earth's surface, a suggested microwave beam would have a maximum intensity at its center, of 23 mW/cm2 (less than 1/4 the solar irradiation constant), and an intensity of less than 1 mW/cm2 outside of the rectenna fence line (the receiver's perimeter).

Some potential solutions to the overcrowding of the GSO problem include reforming the legal regime in space, allowing appropriation of GSO slots, or at least allowing developed nations to use the slots for SBSP satellites on a first-come, first-serve basis. These changes are warranted because space technology has changed so much since the passage of the Outer Space Treaty of 1967, and SBSP satellites also offer a highly efficient solution to the increased demand for clean energy due to population growth.154 But there is another solution to our energy crisis that still involves using SBSP satellites and bypasses the legal hurdles that the GSO poses. While most studies of space SBSP satellites have focused on systems that require a satellite to be stationed in GSO, a lesser-known design promises to reduce both costs and political drama by placing the satellite into a highly elliptical Lower Earth Orbit ("LEO") around the Earth.155

When SBSP satellites were originally designed, they were very large and required a two-stage Earth-to-orbit transportation system in which hundreds of astronauts assembled the satellite in space before sending it into GSO.156

The original satellite design was prohibitively costly and large. The problem with launching SBSP satellites into GSO is that the GSO is approximately thirty six thousand kilometers away, so SBSP systems require satellites with large microwave transmitters on the satellite and receiving stations on Earth.157 Decreasing the size of the solar power satellite and placing it into an orbit closer to Earth would reduce construction and launch costs and allow for a smaller receiving antenna ("rectenna") to be built on Earth.158 More importantly, if a solar power satellite could be placed into an LEO and function efficiently, countries would not be concerned with obtaining the right to own or access a GSO slot or with interference that may result to neighboring telecommunications satellites. There is currently no system of registration or allotment that limits access to LEO/MEO.

One of the few benefits of placing a satellite in GSO, on the other hand, is that GSO satellites stay over one area of Earth's surface. LEO's elongated, elliptical orbit is advantageous because long dwell times over the receiver during the approach to and descent from the apogee allow the satellite to transmit signals to the ground receiver in a manner similar to SBSP satellites beaming from GSO.160 SBSP satellites could operate from a three-hour LEO, where SBSP satellites would slow down near the apogee and speed up near the perigee.161 A three-hour LEO would provide a dwell time of about two hours over the receiving station on Earth; satellites in this orbit would orbit the Earth eight times in a twenty-four hour period.162

SBSP satellites placed into LEO Molniya orbit will only be able to beam seventy percent of the power that they collect because they will be out of contact with the rectenna when they travel in the perigee in contrast, satellites in GSO are in constant contact with the rectenna.164 Yet satellites placed into LEO Molniya orbit are also much closer to Earth than those placed in GSO, so both the satellites and the receiving stations can be made significantly smaller.165

Since the mass of the solar power satellite is directly proportional to the distance between the transmitter and receiver, LEO satellites can be much smaller than the size of GSO-based SBSP satellites, and LEO rectenna can be much smaller than GSO based rectenna. Consequently, SBSP satellites placed into LEO will be much more affordable than large GSO satellite systems. In sum, a solar power satellite Molniya orbit system based around the LEO would be much more economically and legally viable because LEO satellites are cheaper and do not require launching countries to use a limited slot in outer space.

The most defined way to allow SBSP satellites access to the GSO would be to establish an agency similar to the ITU to manage access to the GSO for energy purposes.197 Nevertheless, this would only be a temporary solution because, even if countries had rights to slots in the GSO, many countries would not have the capacity to accommodate many satellites.

A more viable, permanent solution that allows all countries equal access to outer space side steps the legal issue of property rights altogether. Instead of placing SBSP satellites into GSO, launching countries could position SBSP satellites in a highly elliptical LEO Molyina orbit.198 Although such satellites would not be in constant contact with the receiving station on Earth, the satellites and stations could be made significantly smaller and cheaper.199 More importantly, satellites in LEO do not require assigned slots.

A SBSP system based in LEO Molyina orbit would allow significantly more countries to have access to SBSP, without the need to reform property rights in space or to establish a new agency to manage access to the an orbit in space.201 An LEO SBSP system is more economically and legally viable than any system based in the GSO because LEO satellites are cheaper and do not require a highly desired slot in outer space. SBSP satellites can provide highly efficient, clean energy to the world. Launching countries can take advantage of SBSP satellites by placing them into a highly elliptical LEO Molyina orbit where, unlike GSO, there is no controversy over legal ownership of outer space.202

A new baseline model is needed that substantially reduces mass to orbit. The proposed 3-hour sub-Molyina orbit SBSP satellite could provide a new and better model. This orbit will be especially useful in providing power to the higher latitude markets of Canada, Russia, US-Alaska and Europe.

The primary challenge for space solar power towers is economics. Over half the cost of SPS is associated with launch costs. To reduce launch costs the size of the system must be reduced. This proposal reduces the system mass substantially. The sub-Molniya orbit places the satellites closer to Earth and allows for their servicing. *J. E. Drummond* notes that at + 64.4 degrees these two orbits alone would be adequate to supply the base load needs of centers between latitudes 40 and 60 degrees with rectenna an order of magnitude smaller than those required to receive power from an antenna of given area at geostationary orbit.

I first looked at Equatorial low Earth orbits (ELEO) and decided there were too many problems, such as Earth shading and short transmissions times (the transmission problem was solved later), and looked for a better solution. The solution seemed to be a low Molniya orbit. The Molniya orbit was discovered by the Soviets and used for communications and spy satellites. Today the Sirius Satellite Radio satellites use this orbit to transmit their signals to US customers. A Molniya orbit is simply a stable elliptical orbit with an inclination of 63.4 degrees. This inclination puts the satellite high above the Earth and looking down at the northern industrialized countries.

Due to the ellipse the satellite would have a long dwell time over the ground receiver (rectenna). Here the satellites have high apogees and low perigees. The drawback of using a low Molniya orbit is that you only get to utilize the satellite for percentage of the orbit. However, the mass reductions in transmitter mass are so great that it is still far more cost effective than GEO based systems and this translates into smaller, cheaper satellites and much lower launch costs.

Also, the ground rectenna is much smaller at just a fraction of the size needed for GEO SBSP. The satellite would orbit the Earth several times each day and could supply power to ground rectenna as the Earth rotates underneath the satellite.

How profitable? For the Molniya orbiting satellites and assuming the satellites will operate at full capacity as peak power systems and that electricity will sell at for 16 cents a kilowatt hour, a 10,000 megawatt satellite constellation will generate $9,450 million a year in revenue (revenue discounted by 30% for lost utilization). This adds up to $283 Billion over a thirty-year period.

These satellites will produce enough revenue to pay off the original investment, including the support systems, and return a very handsome profit. After return of the initial investment, the cost of energy from the satellites will drop to the cost of operating and maintaining them. There is no fuel to buy and no more debt to pay. The benefits will come to all of us in the form of very low-cost energy.

To produce 10,000 megawatts of power we would need 100 satellites each producing 100 megawatts of power. The cost to launch these satellites is estimated at $100 million each assuming the use of the Russian Proton launch vehicle. This gives a total launch cost of $10 billion. Assuming a cost per satellite of $300 million the cost of the constellation would be $30 billion for the satellites. Subtracting those costs from the $283 billion in revenues leaves $243 billion over thirty years or $8.1 billion annually. Clearly the deployment of these satellites into the peak power market would be hugely profitable.

I took this just a bit further and designed a proprietary photovoltaic system for the satellites with an efficiency of 50% based on the "Rainbow" beam splitting concept. This concept uses photovoltaic cells designed for specific bandgaps. This combination of high efficiency, high solar concentration, low orbits and small satellites provides a solution to the cost effective development of SBSP.

With an investment of approximately $1 billion these designs could be orbiting in less than ten years. Note that this one time funding level for development is only half of one year's funding for "clean coal" demonstrations proposed in the US and 1/60th the amount spent each year in the European Union on Cap and Trade.

This estimated investment only covers the development cost as the operational satellites would be leased to Countries, States, Major Cities and Private Energy Companies. This approach provides an almost unlimited access to capital.

It is envisioned that there would be a total of forty rectenna with twenty around the equator and twenty in the north receiving a total of 20,000 megawatts of power from two satellite constellations of 100 satellites each. This does however create a unique problem as there is no launch service provider that can place that many satellites into orbit in a reasonable timeframe without substantial new investment in launch capability. At most we could expect to launch five satellites each year based on existing launch capability of any single provider. Therefore, it would take forty years to complete both constellations without substantial new space lift capability.

Due to the mass to orbit requirements for geosynchronous SPS, other options should be considered. As will be shown, the closer the orbit to the Earth the more efficient space solar power systems can be. But there is a serious limitation. The primary problem with solar power satellites in LEO or MEO is transmission time over the receiver.

We think a solution can be found in the use of Elliptical Orbits. Due to the second Kepler law of planetary motion, the satellite spends about two thirds of the time near its apogee where it provides what is very close to a stationary perspective centered over the high latitudes. A PowerSat operating in a low Molniya orbit can achieve a utilization rate of 70%.

While this is less than the 100% rate of GEO PowerSats, the mass reduction possible by being located closer to earth more than offsets the handicap of reduced transmission time by allowing for satellites that are smaller and lower cost, which also means the launch costs can be less expensive.

Elliptical Orbits:

A highly elliptical orbit (HEO) is characterized by a relatively low-altitude perigee and an extremely high-altitude apogee over Earth. An elongated orbit can have the advantage of long dwell times over the receiver during the approach to and descent from apogee.

Bodies moving through the long apogee dwell can appear still in the sky to the ground when the orbit is at the right inclination, and when the angular velocity of the orbit in the equatorial plane closely matches the rotation of the surface beneath. Elliptical orbits are useful for communications satellites. Sirius Satellite Radio uses HEO orbits to keep two satellites positioned above North America while a third follow-on satellite quickly rounds the southern part of its 24-hour orbit. A solar power satellite placed in a 3 hour elliptical Molniya Orbit would have a utilization rate of 70%. This is a higher percentage of utilization than LEO PowerSats while still being closer and allowing for smaller antenna.

SPS in Elliptical Orbit:

Successful SPS designs will be those that are technically feasible and economically affordable. One way to shorten time-to-term, and thereby alleviate some of these constraints, will be to look for a workable non-GEO orbit. The author suggests a highly elliptic 3-hour orbit Molniya orbit. These satellites will generate zero pollution and zero emission energy.

To further reduce satellite transmitter and receiver mass, the powersats can operate in a 3 hour sub-Molniya orbit. This orbit is preferred because it provides more beam time per ground station than a circular orbit. Because the orbit is elliptical, the satellites slow down near apogee and speed up near perigee. This provides a "hang time" of about 2 hours over the receiver.

As will be shown, the closer the orbit to the Earth the more mass efficient space solar power systems can be. But there is a serious limitation. The primary problem with solar power satellites in LEO or MEO is transmission time over the receiver. A possible solution can be found in the use of Elliptical Orbits. Due to the second Kepler law of planetary motion, the satellite spends about two thirds of the time near its apogee where it provides what is very close to a stationary perspective centered over the high latitudes.

A PowerSat operating in a low Molniya orbit can achieve a utilization rate of 70 percent. While this is less than the 100 percent rate of GEO PowerSats, the mass reduction possible by being located closer to earth offsets the handicap of reduced transmission time by allowing for satellites that are smaller and lower cost, which also means the launch cost can be less expensive.

In 2010 the author suggested a highly elliptic 3-hour orbit similar to the Molniya orbits used by the Soviets only operating much closer to Earth.

> "To further reduce satellite transmitter and receiver mass, the PowerSats can operate in a 3-hour sub-Molniya orbit. This orbit is preferred because it provides more beam time per ground station than a circular orbit. Because the orbit is elliptical, the satellites slow down near apogee and speed up near perigee. This provides a "hang time" of about 2 hours over the receiver. This orbit will provide about 2 hours of transmission time per orbit and 1 hours of non-transmission time. When compared to the GEO location, this

new class of sub-Molniya orbits has the potential to substantially reduce SSP satellite mass."[20]

An SBSP satellite constellation placed in a sub-Molniya orbit would have a transmitter sized based on the apogee of the satellite. For a 3-hour orbit the apogee would be 8,000km.

Molniya Satellite Assumptions:

Since the satellite located in a 3-hour sub-Molniya orbit would be 4.3 times closer than GEO, the transmitter would be 18.49 times smaller (Inverse-square law (4.3x4.3=18.49). Therefore, a 6,500,000kg 1GW GEO transmitter would be reduced to 351,541kg. The total power in the beam would be the same only it would be concentrated onto a smaller spot on the ground. If we wanted to keep the beam density the same we could build 18.49 smaller satellites of 54MW each rather than a large single satellite of 1GW.

Molniya ISC 2013: 5.8Ghz 2000x Concentration, 40% efficient PV, 1GW, Mass:

1,822,003kg, Transmitter 457,003kg, Power system 1,365,000kg

/ 18.49 = Transmitter 351,541kg + 30% for power loss = 457,003kg

Power System 1,050,000kg + 30% (315,000) = 1,365,000

To achieve progress in SBSP satellite design, we need to create a new Reference Standard based on lower cost MEO satellite placement.

---

[20] Jones,R (Year).Alternative Orbits for Space Solar Power.Online Journal of Space Communication, volume16

The suggested model would be:

> 3-Hour Elliptical Orbit
>
> Beam time 2 hours per orbit
>
> Orbit Non-transmission Time 1 hour
>
> Utilization Rate 70% (without power storage)
>
> Orbits per day 8
>
> Total beam time 2 x 8 = 16 hours

SS-O Sub-Molniya:

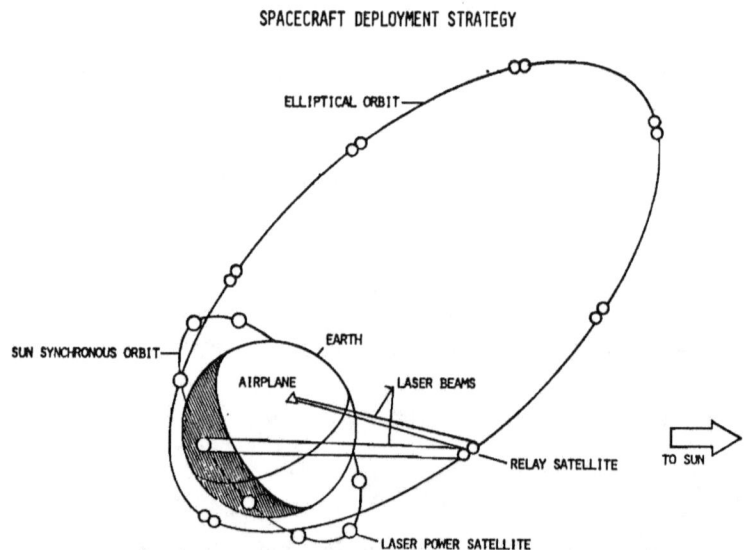

Given the physics of wireless power transmission, when compared to geosynchronous (GEO) orbit at 36,000 km, medium (MEO) Earth orbits located at 10,000 km or less should permit considerable reductions in the size of both the solar power transmitter and the ground receiver. Furthermore, a smaller ground receiver is better suited to servicing such high-density markets as exist in Japan and Western Europe.

The SS-O Sub-Molniya concept was originally proposed as a way to power laser powered aircraft. The PowerSats in SS-O can produce constant power. The reflectors would have a 70% utilization rate. The reflectors would reach a distance of about 12,000km at apogee. This concept would provide good coverage for the Northern Hemisphere.

There are potential variations to these concepts, such as, beam regeneration by refocusing the beam at the reflector and using active repeaters, were the energy is received and retransmitted. Beam reforming would be very energy efficient but would require more complex reflectors. Beam regeneration would have power losses to retransmit the beam. In cases where the beam is refocused or regenerated the distance is measured from the point of refocus or regeneration. In space to space transmission using beam regeneration the beam frequency between the satellites can be different than that needed to penetrate the Earth's atmosphere.

The most efficient of these is the SPR with configurable reflectors. In this concept the energy is produced by a PowerSat is SS-O and the energy is transmitted to the reflector using microwaves or laser. The configurable reflector would refocus the beam and bounce it down to the rectenna of the ground. The power loss using a microwave reflector would be less than 1% and the beam distance from reflector to ground would only be 4,000km when directly above the rectenna.

Beam distance from the PowerSat to the reflector would vary with the closest point being about 3,200km between the two systems and increasing to about 8,000km when the PowerSat is at the North or South Pole. Due to the mass-to-orbit requirements for geosynchronous SPS, other options should be considered. As will be shown, the closer the orbit to the Earth the more efficient space solar power systems can be.

But there is a serious limitation. One problem with solar power satellites in LEO or MEO is transmission time over the receiver. We think a solution can be found in the use of Elliptical Orbits. Due to the second Kepler law of planetary motion, the satellite spends about two thirds of the time near its apogee where it provides what is very close to a stationary perspective centered over the high latitudes. A powersat operating in a low Molniya orbit can achieve a utilization rate of 70 percent. While this is less than the 100 percent rate of GEO powersats, the mass reduction possible by being located closer to earth more than offsets the handicap of reduced transmission time by allowing for satellites that are smaller and lower cost, which also means the launch costs, can be less expensive.

If the technical, economic and societal viability of MEO systems can be demonstrated, then space-based solar power systems in LEO could also prove to be of major interest, when the satellites are as close as the International Space Station more or less 600 km above Earth. The first step in demonstrating either of these possibilities is to move away from past concepts based on solar power satellites stationed 36,000 km from Earth.

There is no logic to placing SBSP Satellites in GEO. The only advantage is the ability to constantly remain over one area on the Earth's surface. Clearly, this advantage is not worth the massive increase in satellite mass that this single advantage gives. For example, the transmitter mass difference between GEO 24-hour orbit and a circular 12-hour orbit is 50 percent. The mass of the transmitter is cut in half simply by moving it closer to Earth, the ground receiver is half the size and the system is more economically viable. To deliver energy to ground receivers 24 hours per day, the design calls for two equal satellite systems spaced 12 hours apart providing coverage to two ground stations (Figure 6). Each satellite will host a transmitter one-fourth the size of the GEO system.

To further reduce satellite transmitter and receiver mass, the PowerSats can operate in a 3-hour sub-Molniya orbit. This orbit is preferred because it provides more beam time per ground station than a circular orbit. Because the orbit is elliptical, the satellites slow down near apogee and speed up near perigee. This provides a "hang time" of about 2 hours over the receiver.

New Reference Design:

To achieve progress in SBSP satellite design, we need to create a new Reference Standard based on lower cost LEO and MEO satellite placement. The suggested model would be:

- 3-Hour Elliptical Orbit
- Beam time 2 hours per orbit
- Orbit Non-transmission Time 1 hour
- Utilization Rate 70% (without power storage)
- Orbits per day 8
- Total beam time 2 x 8 = 16 hours

The highest priority research areas for solar power satellites are those where major improvement can be made in the technical feasibility and cost of the system. The advantages of space-based solar power cannot be realized in the near-term due to the presumed cost of transmitting power from orbit to receiving stations on Earth. These two components are interdependent due to the need for high efficiency power transmission. Since SBSP microwave transmitter size and mass is a direct function of distance between transmitter and receiver, only sub-GEO satellites should be considered. This can be a shocking revelation for people who have always taken it as a given that SBSP satellites must be positioned in a geosynchronous orbit.

Transmitter mass is a large part of SBSP satellite design and therefore affects the system cost, especially launch costs. Since the power transmission subsystem represents about half the capital cost of the total SPS reference system, it is worthwhile to consider the lower orbit alternatives so the technological, environmental, social and political problems and relative advantages may be assessed in comparison with those of geosynchronous forms.

J. E. Drummond notes that at + 64.4 degrees these two orbits alone would be adequate to supply the base load needs of centers between latitudes 40 and 60 degrees with rectenna an order of magnitude smaller than those required to receive power from an antenna of given area at geostationary orbit." (Drummond, 1980)

MEO the Sweet Spot (10,300km) :

Let's consider the Sweet Spot and see what happens there. The Sweet Spot is a 6-hour equatorial orbit at 10,300km located between the inner and outer radiation belts. A 6-hour orbit with three satellites would beam for 2 hours per rectenna providing almost constant power (12% loss during the middle of the night).

Orbit	Distance km	Solar	PV Mass	Freq	Tx Mass	Total Mass
GEO	36,000	1	12,120,000	2.4	3,000,000	25,120,000
GEO	36,000	2,000	1,212,000	5.8	6,500,000	7,712,000
MEO	10,300	2,000	1,212,000	5.8	541,667	1,753,667

Again the mass savings from the closer transmitter are greater than the additional mass due to shadowing.

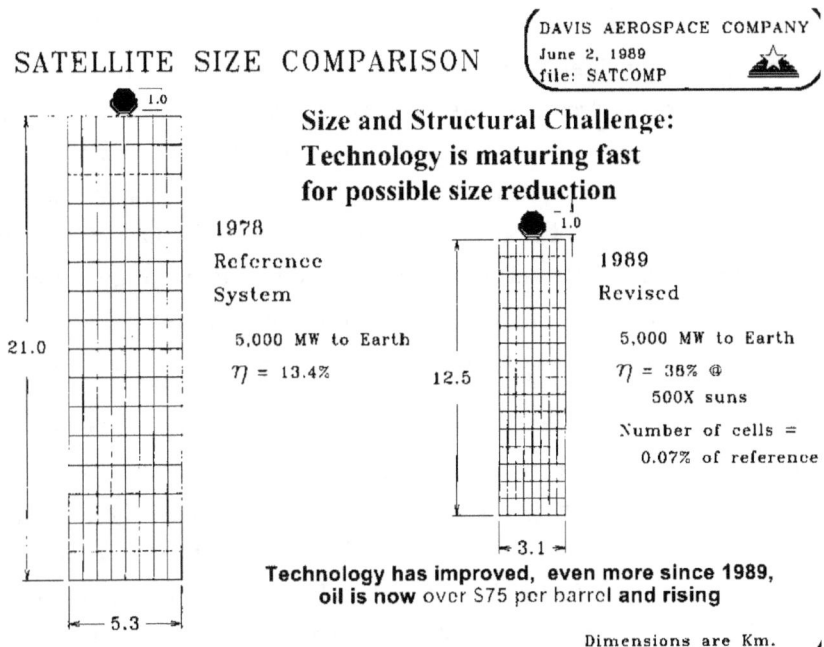

Notice how advances in technology allowed the PowerSat to be reduced in mass using 500x solar concentration the transmitters are still the same size and mass. A better layout, such as the ISC concept would further reduce the mass by removing more structural and wiring mass. By moving to an even higher solar concentration such as 2000x we can make the PowerSat even smaller.

MEO SS-O:

The orbital location of PowerSats plays a critical role in determining the mass of the solar power satellite (PowerSat) transmitter and the size of the rectenna on the Earth's surface. These in turn play an important role in the cost of deploying the PowerSat, especially the cost of launching the PowerSat into orbit as the transmitter makes up a large part of the PowerSats mass.

Here we will consider a new approach to PowerSat orbital positioning by considering a circular sun-synchronous orbit at 5,185.3 kilometers with an inclination of 142.1 degrees. Locating the PowerSat at this location offers several benefits and only one major drawback. The benefits include small transmitter, small rectenna, small minimum power levels, constant energy production and constant energy delivery. The one drawback in that this location is in the inner Van Allen radiation belt, which is a high radiation environment.

Locating the PowerSat in the radiation belt is problematic as the radiation can damage the solar cells. However, the use of highly concentrated solar energy can potentially be used to heal the damaged solar cells, thereby providing a reasonable solution to the problem of radiation damage. Ideally the use of highly concentrated solar energy would be used in the design of the PowerSat anyway to reduce the mass to orbit requirements.

Concentrated solar energy can potentially reduce PowerSat energy production system mass by an order of magnitude (90%). This combined with a close orbit which can reduce transmitter mass by an order of magnitude (90%), will allow for the development and deployment of much smaller, more mass efficient PowerSats.

In addition to these advantages the PowerSats will also have a much lower minimum power level which means that much smaller systems can be deployed incrementally rather than building massive PowerSats as has been proposed in the past. This research paper will detail the orbit mechanics and benefits of PowerSats located in a circular sun-synchronous orbit at 5,185.3 kilometers with an inclination of 142.1 degrees and utilizing concentrated solar energy at 2,000 suns concentration for low mass and radiation healing.

Electric Space: Space-based Solar Power Technologies & Applications

By placing the PowerSat constellation in a medium Earth orbit (MEO) that is sun-synchronous the constellation can provide continuous power to specific sites on the ground. However, SPS in this orbit, which is the most intense region of the Earth's inner Van Allen radiation belt, would require dramatic advances in radiation hardening for all systems. Solar arrays, in particular have been found to be susceptible to degradation due to exposure to radiation. Self regeneration of solar cells using heat has already been demonstrated at the experimental stage and the high heat level available at 2,000 suns solar concentration should be sufficient to heal the cells.

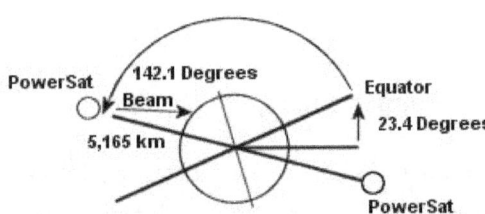

Sun-sychronous orbiting PowerSats at 5,165 km/142.1 degrees

Orbital Data

Orbital height	km	5,165.3
Angle off sun		56.5
Orbital speed	km/h	21,160.9
Orbital period	hours	3.4254
Orbits/day		7
Radius		1.80 Earth radius
Orbit Circumference km		72054.31

Transmitter Size Estimate at Different Frequencies

max sep	5165	km	beam half			Tx diameter	
Freq	Rx Diameter		angle	Radians		km	metres
2.5 GHz	5.11	km	a/2 =	0.000989351	d =	0.147976	147.9757
5.8 GHz	3.35	km	a/2 =	0.000648596	d =	0.097292	97.29233
15GHz	2	km	a/2 =	0.000387222	d =	0.024874	24.87355
38 GHz	1.5	km	a/2 =	0.000290416	d =	0.033165	33.16474
1.5 mcron	1.00E-02	km	a/2 =	1.94E-06	d =	9.45E-04	0.945195

PowerSat Size Comparison :

NASA 1980 Option 1: 5GW, 1x Concentration, 16% efficient PV, 5GW, Mass 51,000,000kg, Transmitter 13,000,000kg, Power 38,000,000kg

First let's deal with power production and then we will deal with the transmitter element. The old 1980s design used 16% efficient solar cells in a flat panel design and had a solar array mass of around 38 million kilograms. If we increase the efficiency to 50% using spectrum beam splitting we could reduce the mass by 3.125 times, which would reduce the mass from 38 million kilograms to 12.16 million kilograms.

By moving to a design using low mass inflatable reflectors for concentrating the suns energy at 2,000 suns we can reduce that further by approximately 90% or 1,216,000 kilograms. A typical system as proposed for SBSP might have large diameter parabolic reflectors of hundreds of square meters, each focusing sunlight on small receivers of PV cells. An advantage is that the area of expensive PV cells is minimized and concentrated sunlight increases cell efficiency, but solar energy not converted to electricity must be dissipated.

Now for the transmitter, the 1980s design used a transmitter operating at 2.45 GHz with a mass of around 13 million kilograms. This was later moved to 5.8 GHz which cut the transmitter mass in half to about 6.5 million kilograms. Now let's integrate our PowerSat into SS-O at 5,100kilometers. This is a much shorter beam distance than GEO and would we could build both smaller transmitters and smaller Earth rectenna. With a beam distance of 5,100 kilometers we are 7 times closer than GEO and the beam spread would be 49 times less. So now we can make our PowerSat transmitter much smaller and less massive at about 132,653 kilograms.

Now our PowerSat has a total mass of 1,348,653 kilograms. The beam is now 49 times more concentrated than before which is too much energy for the rectenna. To keep the energy level the same watts per square meter as before we can break the PowerSat into smaller PowerSats. This would give us PowerSats that are 49 times smaller than the GEO PowerSats, each with a mass of approximately 27,523 kilograms. Each PowerSat would deliver 106,382,978 watts of energy.

Note that these PowerSats can be launched into orbit using several existing launch vehicles; therefore no new launch vehicle funding is required. We can see that the combination of a closer orbital position combined with highly concentrated solar energy can lead to massively smaller, more affordable and more easily deployed PowerSats.

Each PowerSat will pass over each rectenna seven times per day delivering a steady stream of energy to four Earth based rectenna.

Launch Costs:

To obtain 5GW of energy by launching 49 PowerSat and assuming a cost of $100 million per launch would cost approximately $4.9 billion. Space Launch System (SL): Note the huge size of the vehicles tanks. SLS can lift approximately 80,000 kilograms into orbit.

NASA's Space Launch System could deliver three multi-megawatt PowerSats into low Earth orbit in a single mission. Such PowerSats could beam energy to several 106,382,978 megawatt space tugs traveling back and forth between the Earth and the Moon.

This would provide a low cost in-space transportation to support Lunar or free space colony settlement. Such a system could also be used as a booster system for crew and cargo transport to Mars. The basic technology already exists and this technology when combined with the innovation offered by the authors makes SSP an economically attractive technology for near future energy production. In the future new heavy lift launch vehicles, such as SLS could deploy ever larger versions of these PowerSats, possibly of gigawatt scale as technology reduces the mass of the PowerSats.

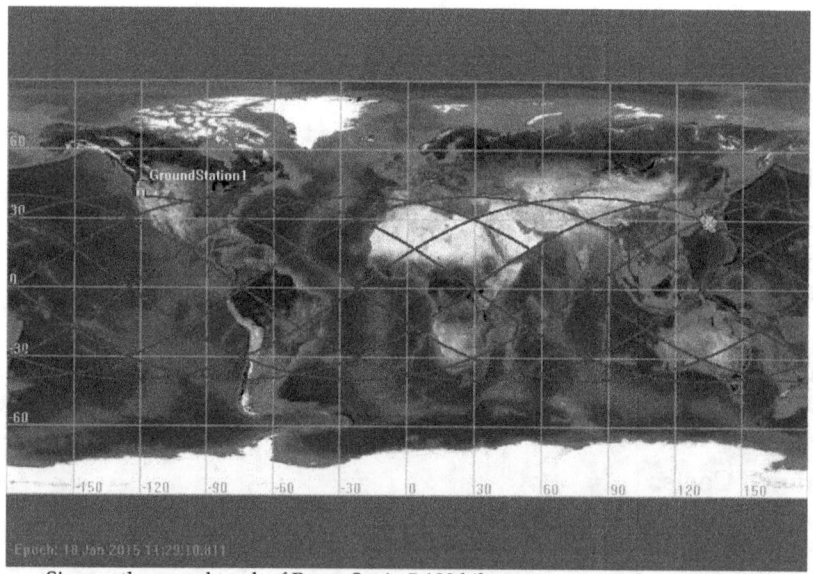

Six month ground track of PowerSat in 5,100 kilometer sun-synchronous orbit.[21]

---

[21] Image credit Gary Snyder

Chapter Eight

The Space Grid

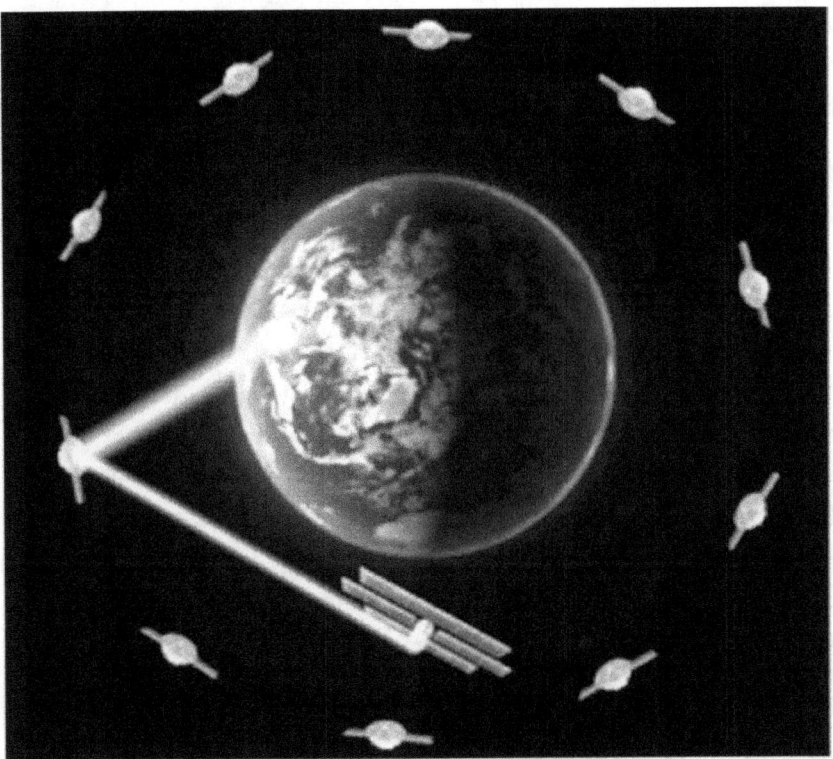

There have been a number of studies on the economics of space-based solar power (SBSP); however, these studies have always focused on locating the solar power satellite in GEO. Medium Earth Orbits (MEO) and Low Earth Orbits (LEO) are sometimes mentioned in these studies but the discussion always focuses on GEO. This is interesting because the limitation on successful deployment of SBSP systems has always been limited by the mass to orbit problem, a problem which could potentially be solved by MEO and LEO satellite positioning.

It has been known since at least 1980 (J. Drummond, 1980) that positioning the power satellite in a lower orbit would reduce the mass of the satellite transmitter by an "order of magnitude". Since the transmitter makes up a large part of SBSP satellite mass a 90 percent reduction in transmitter mass would seem to be very beneficial. There are however problems with MEO and LEO positioning and the biggest problem is Earth Shadowing. However, even this problem has a solution as demonstrated by The Space Grid concept proposed by the Jones, R.

The Space Grid is a concept invented by Jones, R. in 2010 as a way to solve the problems related to LEO PowerSats. While the concept is new and requires further study it does provide another alternative to the distance problem related to GEO PowerSats while at the same time providing solutions to the shadow problem, ocean problem and beam time problem related to LEO PowerSats. The Space Grid allows the PowerSat to be located in LEO thus removing the in-space transportation requirement posed by GEO while being able to provide constant power. Additionally, it does not have the large angle beaming problems posed by MEO PowerSats.

In comparison with proposed alternatives, the Space Grid approach has clear potential to enable radical improvement in terms of higher performance, lower cost, less mass, higher reliability, improved safety, and ease of manufacturing and maintenance. When combined with today's level of technology such as Massive Solar Concentration and Spectrum Beam Splitting for high efficiency, it seems to provide a logical solution to PowerSat development.

The Space Grid relies on two concepts; sun- synchronous orbits (SS-O) and a space power relay (SPR). Let's take a quick look at these two concepts and then a closer look at how the Space Grid uses them.

Sun-synchronous orbits:

Sun-synchronous orbit (SS-O) is a special case of the polar orbit. Like a polar orbit, the satellite travels from the north to the south poles as the Earth turns below it. The orbital plane of a sun-synchronous orbit must also precess (rotate) approximately one degree each day, eastward, to keep pace with the Earth's revolution around the sun. Sun-synchronous orbits are typically low Earth orbits (LEO) ranging from 550 to 850 km and are therefore close to the surface of the Earth. There is a special kind of sun-synchronous orbit called a dawn-to-dusk orbit. In a dawn-to-dusk orbit, the satellite trails the Earth's shadow. When the sun shines on one side of the Earth, it casts a shadow on the opposite side of the Earth. Because the satellite never moves into this shadow, the sun's light is always on it. Since the satellite is close to the shadow, the part of the Earth the satellite is directly above is always at sunset or sunrise. This allows the satellite to always have its solar panels in the sun.

Satellite in Sun-sync dusk to dawn orbit. Note that the satellite never enters the Earth's shadow and the solar panels can always face the sun.

Another advantage of SS-O for PowerSat stationing is that they can always be facing the sun side of the Earth. This means that all of the power transmitters can be located on the day side where power is need most. A GEO PowerSat will follow the ground transmitter day and night and does not have the flexibility of the SS-O PowerSat without adding space reflectors.

This would defeat the primary reason for considering GEO. Since the goal is to deliver as much power as possible to where it is needed the most, the Space Grid would transmit all of the power to the day time side of the Earth. Beaming all the power to the day side of the Earth is more productive than Earth based solar power since the solar energy in space is much greater than on the ground. Additionally, the Earth is rotating so the "day side" is always changing but the energy being transmitted by the SBSP satellite is constant.

The Earth's oblateness (J2 parameter in the gravitational model) produces a shift in the orbit. This will twist the orbit plane around so it passes the equator at a different point (node) along the equator (Regression of nodes = useful for the sun-synchronous part later) and also push the low point (periapsis) and hight point (apoapsis) further along the orbit (line of apsides).

The line of apsides motion varies depending on inclination. For this reason an elliptical orbit would move the low points around and not be low over a specific region of the earth. While at certain inclinations and altitudes (semi-major axis or period) such as 63.4 degrees and a half siderial day, have a stable periapsis they won't be sun-synchronous (but star-synchronous!). Therefore, we will use a circular orbit.

Sun-synchronous: This is to say the satellite orbits in such a way that the equator crossing is always at the same time of day, or solar ground angle. This is useful for earth observing satellites to get the same shadows in their images.

A normal satellite flying west to east, and inclined to fly above and below the equator will have its orbital plane pushed westward. In effect rising earlier each day than just attributed to the earth's revolution around the sun. By launching into a backwards orbit, e.g. opposite of the earth's rotation (spin), the rotation of the orbital plane can coincide with the sun's position changing due to the Earth's revolution. In effect it needs to process 360 degrees backwards in 356.24 days.

The governing equations are:

Inclination = $ACOS(-2*P*sqrt(Rsat^7/MuEarth)/3*J)$

Rsat = The orbital radius of the satellite = altitude + 6378.141 Km

P = solar period = $2(Pi)/(365.24*24*60*60)$ Radians/sec

MuEarth = GMearth = 3.986×10^5 km^3/sec^2

J = the J2 oblateness factor of the earth = 1.755×10^{10} km^5/Sec^2

Rsat = Satellite Altitude + Rearth

Rearth = 6378.141 km

Note there isn't enough influence from the J2 effect to precess the orbit to track the sun when the orbit is at or above 5974 km altitude (which is nearly flat to the equator in retrograde).

A Repeat track on surface of the earth: This is not the same as sun synchronous, but refers to the concept that at the same point on the earth should be overflown each day. Many earth observing satellites in Sun-Synchronous orbits will re-fly over the same position at the same time of day at some other schedule, like every 6 weeks. To get daily repeats, pretty much dictates the orbital period shall be an even division of 24 hours. The easiest example is geosynchronous orbit (1 orbit in 24 hours). A counter example is the GPS constellation in nearly 12 hour orbits. It was selected to change by 4 minutes each day for early system testing when there were only a few satellites and they wanted to test at assorted times of the day.

The equations are:

Period = 2(Pi)sqrt(Rsat^3/MuEarth) and

n*period = 24 hours where n is an integer

No shadow: Note that the typical geostationary satellites have two periods of shadow each year when the orbital plane on the equator and the ecliptic plane of the sun cross (similar to lunar eclipse). This is also noticed by satellite TV users when the satellites pass in front of the sun (similar to a lunar eclipse). A sun-synchronous orbit can ovoid these periods of shadow by tracking the sun.

The inclination with respect to the Earth will be fixed by the sun-synchronous requirement, and the Sun angle (ecliptic) is inclined by 23 degrees with respect to the Earth's equator, so the solid angle sum of the factors has to be less than this. Seasons and precise placement, and even tracking of solar arrays need to be taken into account for final selection.

Station-keeping requirements will also depend on perturbations, drag and other needs, but they are affected by specific ground tracks and need to be simulated for specific cases.

Approximate current solutions to all above criteria :

Altitude (Km)	Inclination	Orbits/day	Angle off sun
566.8	97.7	15	23.3
893.	7 99.0	14	28.7
1262.0	100.7	13	33.4
1680.8	103.0	12	37.7
2162.1	106.0	11	41.7
2722.4	110.1	10	45.5
3384.6	116.0	9	49.2
4182.1	125.29	8	52.8
5165.3	142.1	7	56.5
6414	(no synchronous)	6	60.1

Space Power Relay:

Space Power Relay (SPR) has been proposed in the past. (N. Komerath, N. Boechler, S. Wanis R. Dickinson and J. Mankins, D. Criswell D Ehricke, K., N. Komerath, N. Boechler I. Bekey, R. Boudreault. The SPR proposed by Ehricke placed the reflector in GEO so the power beam had to travel 36,000 km to the reflector and then another 36,000km back to Earth. In the proposed system the space-based reflector was estimated at 2,500 tons per million kilowatts. At an estimated 2,500 tons per million kilowatts, the space reflector was considerably lower in mass than the SBSP designs of the day, which were estimated at 8,000 tons per million kilowatts received at the ground rectenna.

Ehricke suggested that the reflector could be placed into a lower orbit to reduce transmitter size to 1 million kilowatts and rectenna size from 55 square miles to 15 square miles. There have even been proposals for Moon to Earth SPR systems using space-based reflectors to move the energy around the Earth (*Criswell (9)*).

A similar concept was proposed by the Federal Aviation Administration (FAA) in the 1960s for air traffic control. In that system, the microwave beam generated on the Earth's surface would bounce off a large space reflector in GEO and then back to the Earth to track moving aircraft (*Grumman*).

The FAA concept was not proposed as a power relay; however many of the components were the same as those required by space power relay. The problems with past power relay concepts for a Space Power Grid (SPG) are the large mass requirements. Although the SPG is much lower mass than SBSP the mass requirements for past concepts are still large when the system are located in GEO, or on the Lunar surface.

It has been suggested by N. Komerath that SBSP could be added to a SPG system. They proposed a space power grid (SPG) that beamed energy into space from collection locations on Earth and bouncing it off lightweight reflectors to earth-based microwave collectors for local distribution. In the SPR concept, the satellites act as waveguides, and do not perform conversion to DC, therefore it is very efficient.

How the Space Grid works:

Placing a SBSP satellite in SS-O can be very beneficial as it would allow the satellite to produce constant power. By using SS-O orbiting PowerSats, these could produce the same power levels of PowerSats located at geostationary orbits, as it was originally proposed by Peter Glaser in 1969 (1) and which is still discussed by many in the space community. Moreover, by locating a PowerSat in the SS-O (2722km) it would be much closer to Earth than a PowerSat on GEO (36,000km), therefore allowing for the use of much smaller transmitters and rectenna (Drummond 2, Jones 3) and so consequently reducing the amount of mass to be transported into the operational orbit.

Combining SS-O PowerSats using WPT with SPR for power distribution would reduce the minimum size of the PowerSat by 250% compared to a GEO PowerSat. The minimum power level of an SBSP satellite in GEO is approximately five GW. This is due of the great distance of 36,000 km the microwave beam has to travel to get from GEO to the rectenna on the Earth's surface. As the beam travels this great distance it spreads. In order to have enough power in the beam to activate the rectenna of the ground you have to build a very large PowerSat in GEO.

With the Space Grid the distance traveled is much shorter since beam travels from the PowerSat in an 2,722 km SS-O to the reflector in a 4,000 km equatorial orbit and then to the Earth.

The shortest distance would only be 3,200km when the two satellites line up at the equator over the rectenna and the maximum distance would be about 8,000 km using a constellation of ten reflectors. This is only 38.7% of the distance from GEO. This reduces the minimum size of the PowerSat from 5 GW to only 1.935 GW (rounded to 2GW). Additionally, we will see a 60% decrease in the aperture of the PowerSat transmitter. The proposed solution, which is a new invention, is to place low mass Reflector Satellites in an Equatorial Medium Earth Orbit (EMEO) and use the reflector satellite to reflect the power transmitted from the PowerSat to the rectenna on the Earth's surface.

Placing just a few reflector satellites in EMEO to reflect the energy to rectenna on the Earth's surface is potentially a low cost solution since the reflectors can be very low mass inflatable structures The development of an economically viable space-based solar power (SBSP) system is critical to the Earth's future and for future space development.

PowerSat technology is also critical to supporting sustainable private and government space ventures, including space lift, space exploration and space infrastructure development. Such a system would greatly expand the need for space lift capability from small reusable launch vehicles for SBSP satellite maintenance to large expendable launch vehicles for deploying GW class SBSP satellites into orbit. The technology needed for SBSP is also needed for in-space solar electric transportation systems needed for space colonization as the technology is the same.

The hope has been that gradual improvement in photovoltaic or other technologies such as thermal systems would solve the mass to orbit problem for SBSP systems. However, this in itself does not appear sufficient to make SBSP economically viable. This paper presents a new architectural option for SBSP using a Sun -synchronous orbit (SS-O), wireless power transmission (WPT) and a space power relay (SPR).

This new concept is called The Space Grid. The Space Grid relies on the use of two separate satellite constellations. The power satellite (PowerSat) constellation is placed in SS-O dusk to dawn orbit at 800km and has access to constant sunlight and is used to produce the power. The Equatorial reflector satellite (ReflectorSat) constellation is in a 4,000km equatorial orbit and is used to distribute the power to the rectenna on the Earth's surface.

The power is produced by the PowerSats in SS-O and beamed to the ReflectorSats in equatorial orbit and then bounced to the rectenna on the ground. This combination allows for the production and distribution of power to the Earth's surface without the problems normally associated with non-Geostationary (GEO) PowerSat concepts and without having to place the PowerSats in GEO.

An alternative approach to GEO must be considered. Recent work by the Jones R. has suggested that smaller PowerSats operating in low Earth Orbit (LEO) would be far more economically viable. This concept was proposed as way to reduce PowerSat transmitter mass by using reflectors in equatorial orbit. SS-O orbits might be 800km, 1,500km or 2,722km. The reflectors would have an 80% utilization rate. This new architectural option for SBSP uses a Sun -synchronous orbit (SS-O), wireless power transmission (WPT) and a space power relay (SPR). This new concept is called The Space Grid. The Space Grid relies on the use of two separate satellite constellations.

The power satellite (PowerSat) constellation is placed in SS-O dusk to dawn orbit at 800km and has access to constant sunlight and is used to produce the power. The Equatorial reflector satellite (ReflectorSat) constellation is in a 4,000km equatorial orbit and is used to distribute the power to the rectenna on the Earth's surface. The power is produced by the PowerSats in SS-O and beamed to the ReflectorSats in equatorial orbit and then bounced to the rectenna on the ground.

This combination allows for the production and distribution of power to the Earth's surface without the problems normally associated with non-Geostationary (GEO) PowerSat concepts and without having to place the PowerSats in GEO. The Space Grid reduces the mass of a PowerSat transmitter by approximately 67% by moving it closer then past GEO concepts and allows for higher power levels and therefore much smaller (60%) and less costly rectenna on the ground and reduces the minimum size from 5GW to only 2GW allowing quicker deployment of space energy to solve the Earth's energy problems. WPT transmission could be microwave or laser but for this paper microwave will be used for easier comparison with past concepts.

The space grid approach integrates the issues of global warming and energy demand with the technologies for space-based solar power and space power relay. Combined these two technologies offer a potential solution to an energy hungry planet. When you consider that there are currently plans to build 50 new coal-fired electrical plants across Europe, dozens of new coal burning plants in China and several dozen new coal and natural gas burning plants across the US, the ability to generate clean energy in space and transfer to the Earth can play a major role in reducing Global Warming by reducing or eliminating the need for new CO_2 producing power plants.

There are over 49000 electric power plants in the world, generating a total of 2812 GW. Power needs during emergencies, such as the ones in Japan and New Orleans might be better met by transporting lightweight deployable rectenna to the area. These rectenna are simple in function yet they can provide access to large amount of energy directed it to by the Space Grid. The ability of space solar providers to begin delivering power early in the constellation deployment and be able to incrementally increase constellation size can add to the affordability factor. Following this model, space energy providers don't need to spend hundreds of billions of dollars to build a single massive satellite when smaller systems will serve the purpose.

The development of a Space Grid using SS-O SBSP PowerSats with Equatorial orbiting ReflectorSats appears to offer all of the advantages of GEO PowerSats, including constant power production for base-load energy. Additionally, it appears to offer many advantages over GEO stationing including, no LEO-GEO transportation, ease of maintenance due to closer positioning, much less mass to orbit due to transmitter sizing and the ability to direct more energy to the day side of the Earth.

While operationally the concept is more complex than GEO stationing, this is overshadowed by the large reduction in the minimum size verse GEO. What this means is that you can start producing power in space by launching less massive PowerSats and this can move humanity closer to clean energy from space. It also means that we can power in inner solar system and open space settlement to humanity. Newer concepts for SBSP have substantially reduced the mass of the satellites by using large solar inflatable concentrating reflectors, thereby reducing the mass of the solar cells and supporting structure. The mass of power relay satellites (PRS) can also be reduced using similar large reflector technology. Reflectors satellites can be launched into a 4,000km orbit using a small launch vehicle or an entire constellation could be launched using a single Falcon IX Heavy.

Combining SS-O PowerSats using WPT with SPR for power distribution would reduce the minimum size of the PowerSat by 250% compared to a GEO PowerSat. The minimum power level of an SBSP satellite in GEO has to be approximately five gigawatts (GW). This is due of the great distance of 31,000 km the microwave beam has to travel to get from GEO to the rectenna on the Earth's surface. As the beam travels this great distance it spreads. In order to have enough power in the beam to activate the rectenna of the ground you have to build a very large PowerSat in GEO. With the Space Grid the distance traveled is much shorter since beam travels from the PowerSat in an 800 km SS-O to the reflector in a 4,000 km equatorial orbit and then to the Earth.

The shortest distance would only be 3200km when the two satellites line up at the equator over the rectenna and the maximum distance would be about 12,000 km using a constellation of ten reflectors. This is only 38.7% of the distance from GEO. This reduces the minimum size of the PowerSat from 5 GW to only 1.935 GW (rounded to 2GW). Additionally, we will see a 60% decrease in the aperture of the PowerSat transmitter.

This proposed solution is similar except the power is beamed from the PowerSat in SS-O rather than Earth. The proposed solution, which is a new invention, is to place low mass Reflector Satellites in an Equatorial Medium Earth Orbit (EMEO) and use the reflector satellite to reflect the power transmitted from the PowerSat to the rectenna on the Earth's surface. Placing just a few reflector satellites in EMEO to reflect the energy to rectenna on the Earth's surface is potentially a low cost solution since the reflectors can be very low mass inflatable structures.

Assuming the same technology level as proposed in the 1980 NASA/DOE study with a 5 GW PowerSat with 2x solar concentration and 20% efficient solar cells we see that the total mass of was 34 million kg. Of this the transmitter mass was estimated at about 13 million kg or 38.2% of the total mass.

A 2GW PowerSat in SS-O is only 40% of the GEO PowerSat mass or 13.6 million kg. The transmitter mass of the 2GW system being estimated at 38.2% of the total mass would be about 5,263,200 kg. However since the beam distance has decrease by 60% the transmitter mass can now be reduced by 60%.

This would reduce the transmitter mass to only 3,157,920 kg giving a satellite mass of 11,494,720 kg. Assuming a 20% mass penalty for launching into SS-O due to the greater Delta-v required for launching into a near polar SS-O orbit this would be equal to launching 13,793,664 kg into equatorial orbit for a mass penalty of 2,298,944 kg.

However, there is more to this because you would have to move the GEO satellite from LEO to an altitude of 31,000 km and the SS-O satellite from LEO to only 800 km. Moving a satellite that distance would consume lots of propellant even when using efficient ion drives.

According to the 1980s NASA/DOE study the propellant mass would be 1/10th of payload mass, so to move the satellite to GEO would consume 3.4 million kg of ion propellant. Using today's technology we might get a propellant mass fraction of 20% and this would cut propellant mass in half to 1.7 million kg.

Therefore, the total mass to orbit, not counting any space infrastructure, would be 34 million kg plus 1.7 million kg of propellant for a total mass of 35.7 million kg. By comparison the SS-O mass to orbit for a same 5GW of power would be 11,494,720kg times 2.5 or 28,736,800kg plus launch penalty of 20% or 5,747,360kg plus 50,000 kg for the reflectors in EMEO for a total of 34,534,160kg.

Therefore, the mass difference is only about 1,165,840kg. However, to obtain power from space you only need to deploy an 11,494,720 kg 2GW system rather than a 34,000,000 kg 5GW system and this is the main benefit of the Space Grid. It reduces first cost to power and allows a much smaller SBSP satellites to be deployed on orbit more quickly.

The proposed SBSP system would use eight SBSP satellites, ten reflector satellites and eight Earth rectenna. The reason for ten reflectors is related to the view time of each reflector over a rectenna below, which is about eighteen minutes.

From the SS-Orbit each PowerSat can view most of the reflectors and chose which one to beam to using electronic beam steering. The proposed constellation can provide power to the eight ground stations. With eight SBSP satellites, each producing 2GW of energy the constellation would deliver 16 GW of clean energy to the Earth.

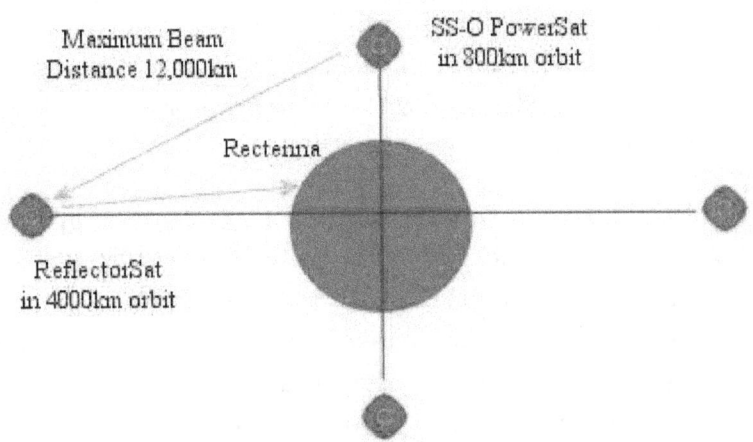

SS-O PowerSat Constellation with two PowerSats and Equatorial ReflectorSat Constellation.

For SBSP, the dimensioning of the RF power transmission system results from an adequate balance between the definition and sizing of the receiver system (rectenna) and the definition of the SBPS transmitting system, the key driver being the transmission frequency. For an SBSP system operating in SS-O and incorporating an SPG in equatorial orbit, the addition of the space-based microwave reflector has to be taken into consideration.

Power loss at the reflector should be less than 1% (4). The much smaller transmission distance, which is only 33% of that of a GEO Powersat, means that both the transmitter and rectenna can be 67% smaller than a GEO based system which increases the economic viability verse concepts based in GEO.

In comparison with proposed alternatives, the Space Grid approach has clear potential to enable radical improvement in terms of higher performance, lower cost, less mass, higher reliability, improved safety, and ease of manufacturing.

Placing a SBSP satellite in SS-O can be very beneficial as it would allow the satellite to produce constant power. By using SS-O dawn-to-dusk orbiting PowerSats, these could produce the same power levels of as those of PowerSats located at geostationary orbits, as it was originally proposed by Peter Glaser in 1969 (1) and which is still discussed by many in the space community.

Moreover, by locating a PowerSat in the SS-O (800 km) it would be much closer to Earth than a PowerSat on GEO (36,000km), therefore allowing for the use of much smaller transmitters and rectenna (*Drummond* 2, *Jones* 3) and so consequently reducing the amount of mass to be transported into the operational orbit. There is a launch mass penalty of approximately 20% for launching into near polar orbits verse launching from the equator.

Therefore, the tradeoff is a lower mass space transmitter verse the reduced mass launched into SS-O. The main argument for locating a PowerSat is GEO is that it can provide constant power to a single location on the ground. However, there are other much closer ways to generate and deliver constant power than GEO.

Closing this distance is important as it has a direct effect on the size and mass of the PowerSat transmitter. There are at least two ways to generate and distribute constant power in using orbits that are much closer than GEO. Both of these concepts would use microwave reflectors in a space power relay (SPR) for power distribution. Both ways use PowerSats located in near polar sun-sync orbits but the power distribution is a bit different.

The first way would be to place the reflectors in 4,000km equatorial orbit and transmit the power to bases +-40 degrees of the equator. The second way would be to put the reflectors in a 4 hour sub-Molniya orbit (64.3 degrees) and transmit the power to locations in the northern latitudes.

Space Grid Constellations:

The proposed SBSP system would use eight SBSP satellites, ten reflector satellites and eight Earth rectenna. The reason for ten reflectors is related to the view time of each reflector over a rectenna below, which is about eighteen minutes.

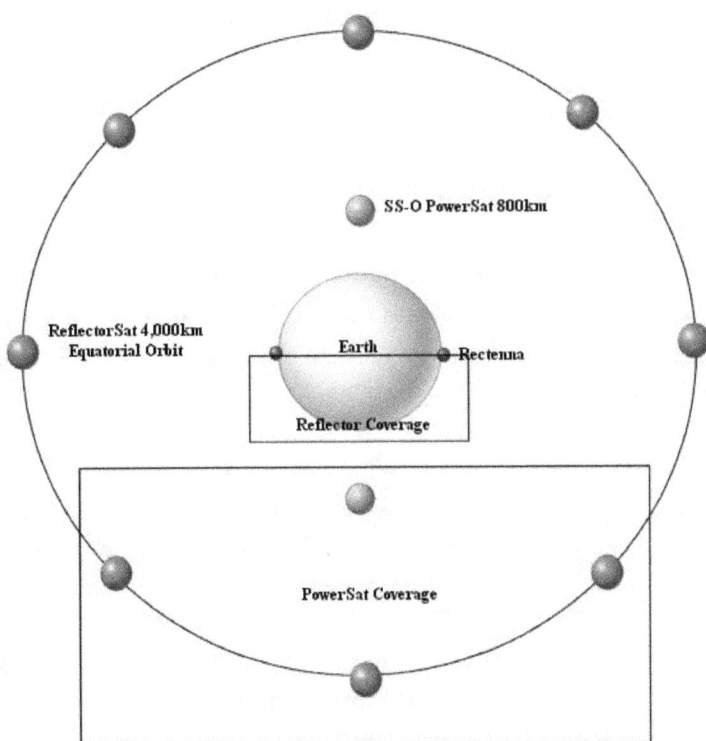

Looking down on the Earth showing the SS-O PowerSat Constellation with two PowerSat satellites and Equatorial ReflectorSat Constellation with eight satellites.

From the SS-Orbit, each PowerSat can view most of the reflectors and chose which one to beam to using electronic beam steering. The proposed constellation can provide power to the eight ground stations. With eight SBSP satellites, each producing 2GW of energy the constellation would deliver 16 GW of clean energy to the Earth.

Newer concepts for SBSP have substantially reduced the mass of the satellites by using large solar inflatable concentrating reflectors, thereby reducing the mass of the solar cells and supporting structure.

The mass of power relay satellites (PRS) can also be reduced using similar large reflector technology. Reflectors satellites with a mass of 5,000kg can be launched into a 4,000km orbit using a small launch vehicle or an entire constellation could be launched using a single Falcon Heavy.

Using the Space Grid for Space Settlement:

Power generation is one of the crucial elements of space vehicles and of future infrastructures on planets and moons. The increased demand for power faces many constraints, in particular the sizing of the power generation system.

The SPS Space Grid system is a candidate solution to deliver power to space vehicles or to elements on planetary surfaces. Beaming energy to spacecraft could lower spacecraft mass and improve mission-economic potential.

A beamed solar electric power (BSEP) system with Sun-synchronous orbiting (SS-O) PowerSats (3,000 km) beaming to equatorial Ion Spaceships would reduces the ships' mass by a factor of 30 compared to chemical, direct drive solar or nuclear ships. This promises a significant reduction in the cost of space transportation.

The technology required for space-based solar power (SBSP) can also be applied to other space applications at much smaller scale, such as Space Tugs, Space Ferries, Spaceships, Space Stations and Planetary Power for the Moon, Mars, etc.

Reusable in-space transportation systems must be capable of both high fuel efficiency and high utilization of capacity, or economic costs will remain unacceptably high.

BSEP systems can provide high fuel efficiency and with enough power high thrust to support cargo and crewed missions. The major contribution of beamed power to the development of space is its unique ability to transfer energy across long distances and across large differences in gravitational potential.

Tugs have the potential to support a wide variety of space mission types, such as retiring geostationary communications satellites, moving cargo to space stations, Moon, Mars and other destinations, cleaning up space debris or on-orbit assembly tasks.

A key feature of the concept is that the Tug is fully reusable and will deliver the cost benefits of reusability verse expendable systems. The Tug is launched just once and remains in orbit throughout its lifetime, where it is never subjected to the risk and stress of repeated reentry and re-launch. This eliminates the recurring cost of subsystem checkout, refurbishment, and recertification, required for concepts that must repeatedly survive the stress of launch and reentry.

This paper will discuss a Beamed Energy In-Space Transportation System or Space Tug using small orbiting power satellites (PowerSats) located in a unique orbit and beaming microwave energy to the orbiting space tug.

Due to the PowerSat location in the sun-sync orbit two of them can provide constant power to up to two space tugs at the same time. The space tugs are hybrid because they have both a wireless power transmission (WPT) rectenna and 0.25 MW of on-board direct drive solar. The energy will power a 0.75 MW Ion Drive Propulsion system.

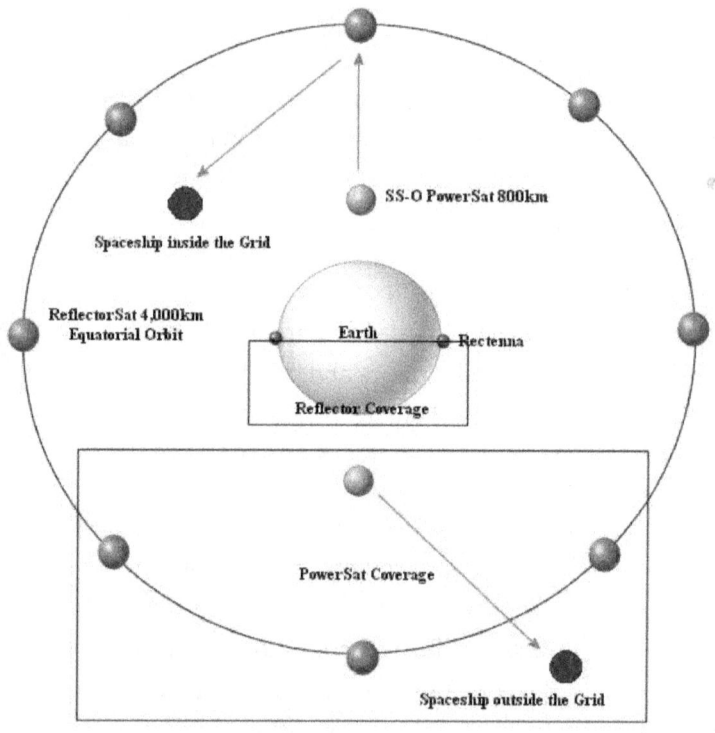

Powering an orbital spaceship with the Space Grid showing one spaceship operating inside the grid and another spaceship operating outside the grid.

Chapter Nine

Solar Electric Propulsion (SEP)

This image of a xenon ion engine, photographed through a port of the vacuum chamber where it was being tested at NASA's Jet Propulsion Laboratory, shows the faint blue glow of charged atoms being emitted from the engine.

The ion propulsion engine is the first non-chemical propulsion to be used as the primary means of propelling a spacecraft. The first flight in NASA's New Millennium Program, Deep Space 1 is designed to validate 12 new technologies for scientific space missions of the next century. Ion propulsion was first proposed in the 1950s and NASA performed experiments on this highly efficient propulsion system in the 1960s, but it was not used aboard an American spacecraft until the 1990s.

The basic technology for SBSP and Solar Electric Propulsion (SEP) have important similarities. This is especially true when considering beamed solar Electric propulsion (BSEP). Multiple studies have shown powerful advantages in employing high-power Solar Electric Propulsion (SEP).

The major drawback of these designs are the low thrust to weight ratios, meaning their acceleration is weak and they are incapable of escaping Earth's gravitational field or moving large payloads. This is offset, however, by the performance of the engine once in space, which can leave the chemical rockets far behind due to its far superior specific impulse.

In other words although the chemical rockets have high thrust to weight ratios the propellant is quickly exhausted at a low specific impulse. The electric propulsion system can run continuously for many months or even years so that, despite the low thrust, they may ultimately build up to a higher total impulse (specific impulse x propellant mass) and hence much greater velocities.

NASA's strategic roadmaps for exploration, science, and advanced technology all consider SEP to be a vital and necessary future capability. Many architecture studies reflect the need for integration, and fueling of these types of missions based on smaller, more affordable launches from the Earth's surface to deliver the required systems and propellants to an assembly point in space.

High specific impulse (ISP) provided by SEP allows the dramatic reduction (approximately 50%) of total launch mass, and hence, a similarly dramatic reduction in the number of launches for such mission architectures.

EP Systems:

An EP system is a set of components arranged so as to eventually convert electrical power from the S/C power system into the kinetic energy of a propellant jet.1 Page 126 Figure shows in schematic form the principal elements of an EP system and its interfaces with other S/C systems.

Typically, the power system supplies regulated dc bus power to a power processor unit (PPU), as well as to other auxiliary elements, such as valves, Received March 3, 1998; revision received June 15, 1998; accepted for publication June 24, 1998. Copyright Q 1998 by the American Institute of Aeronautics and Astronautics, Inc. All rights reserved. *Professor, Department of Aeronautics and Astronautics. †Senior Scientist, Technology Operations Heaters, etc.

The PPU processes this raw power into the specific form required by the thrusters and is usually one of the most complex and challenging EP components, as will be seen in subsequent sections. A regulated, pressure-fed fuel system is shown for illustration, although simple blowdown supplies can sometimes be used. No detail is shown of the plumbing, which often includes series and parallel valves, pyrotechnically opening or closing valves, Pressure Regulator, etc.

The flows to be handled are usually very small, but occur for very prolonged periods of time (months), which presents special challenges for the design of precise flow controllers and leak-free valving.

One of the most important parts is pressure regulators which is acting as a sensor and connects the thruster and PPU to work on time and accurate.

There are a various types to be used, but magnetic MEMS[22] one is invented by *Baghchehsara, A.*[23] in 2010. The system main operation is out of the combustion chamber, and the force transfers between the combustion chamber wall thicknesses with Magnetic waives.

Commands to the various power switches, valves, etc., are supplied by the S/C computer, which also receives and processes a variety of status signals from sensors (only a pair of pressure signals are illustrated). The heart of the system is, of course, the thruster itself, and this paper will concentrate on thrusters. It must be understood, however, that a large proportion of the propulsion engineer's effort must be devoted to the balance of the EP system, which in the end is also usually heavier, bulkier, and more expensive than the thruster(s). Fortunately, aside from the PPU peculiarities, the rest of the system is not drastically different from more familiar cold-gas or monopropellant systems, and indeed, EP has been ted in its gradual introduction from this existing experience base.

Electric Thrusters:

The common feature of all EP schemes is the addition of energy to the working fluid from some electrical source. This has been accomplished, however, in a large variety of physically different devices. Operation can be steady or pulsed; gas acceleration can be thermal, electrostatic, electromagnetic, or mixed; the propellant can be a noble gas, a chemical mono- propellant, or even a solid. Of the many combinations tested over the years, a reduced, but still large number have reached maturity, or are approaching it.

---

[22] Micro Electro Mechanical System

[23] A.Baghchehsara ,Electro-Magnetic Regulator for Liquid Missile Motors, DOI: 10.1111/j.1945-5100.2012.01401_2.x , Volume 47, pages A18-A19 , Wiley Publication, Printed in USA , Aug 2012

Electric thrusters are categorized by type below:

a) Resistojet
b) Arcjets,
c) Hall thrusters
d) ion engines
e) pulsed plasma thrusters
f) field-effect electrostatic propulsion thrusters
g) self-field magnetoplasmadynamic thrusters

Using low-cost high-power SEP propulsion for orbit-raising vehicles

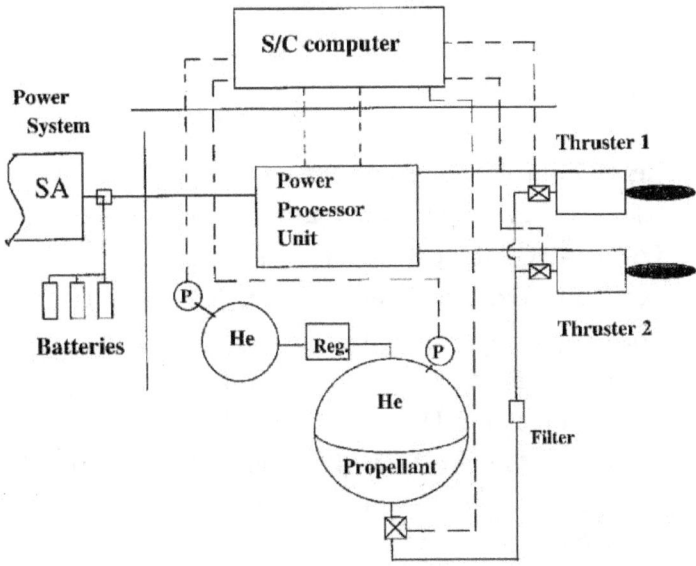

enables the delivery of payloads to low Earth orbit (LEO) via traditional chemical rockets, and then using SEP to spiral those payloads out to higher energy orbits, including Earth-Moon L1, a potential assembly point in space. This approach could facilitate missions to near Earth asteroids (NEAs) and other destinations in deep space.

A flight test of a high-power SEP system will mature advanced in-space propulsion technologies that significantly improve payload mass fraction and can serve as building blocks to higher energy systems in support of human exploration (crew and cargo). A technology flight demonstration will raise the Technology Readiness Levels (TRLs) of SEP system technologies in support of these objectives, as well as guide further technology development for follow-on, higher-power systems.

Solar Electric Space Tug:

- Lunar Ferry Vehicle Initial Mass (includes Module and all Propellant) 35 000 Kg
- Moon base Cargo Module 20 000 Kg
- Xenon Propellant consumed by ion engines 7 300 kg
- S.E.P. Lunar Ferry Dry Mass (Less Cargo) 7 700 kg
- Low Earth Orbit (Leo) Parking Altitude 500 km
- Low Lunar Orbit (Llo) Parking Altitude 100 km
- Total Trip Time (Leo-Llo-Leo) 370 days
- Solar Array Power To Ion Engines 300 kW
- Solar Array Power Density 0.2 kW/m2
- Solar Array Specific Mass (Beginning Of Life - End Of Life) 5 To 6 kg / kW
- Total Ion System Efficiency 0.75
- Ion Engine Specific Impulse 4500s
- Ion Engine Specific Mass 6 kg / kW
- Number of Ion Engines (including spares) 12

Beamed Energy In-Space Transportation System:

While there have been many past concepts for space tugs, sometimes called orbital transfer vehicles (OTV) most all have been based of chemical propellants. Such systems are now outdated because electric space propulsion is far superior.

The propellant needs for chemical tugs are 10-15 times greater than for solar electric tugs and beamed energy space tugs can be 20-30 times lower mass. Both Earth to space and in-space power beaming for space vehicles has been proposed in the past and the technology available for low mass PowerSats has been increasing. We see that the space component, or Space Tug, is very low mass at 42,000kg.

This is a 2 MW sized system using microwave energy beamed from the Earth. Since the beam originates on the Earth the system is limited to 2.4 GHz or with somewhat higher power losses 5.8 GHz because these are best frequencies for beaming through the Earth's atmosphere. This limitation results in greater beam spread over long distances than a space based system which can operate at much higher frequencies. Like solar powered space tugs the amount of energy received when close to the Earth is low even when using four transmitter arrays spaced equally around the Earth because beam time to the space tug is very low until the tug is far from the Earth.

Beamed energy concepts offer an alternative for an advanced low mass propulsion system. The use of a remote power source reduces the weight of the propulsion system in flight and this, combined with the high performance of electric propulsion, provides significant payload gains. There are different types of beamed energy propulsion concepts.

We propose to use a microwave beam generated by Power Satellites (PowerSats) located in a sun-synchronous orbit (SS-O) at 2,722 kilometers (km) around the Earth. By locating the PowerSats in this orbit they can produce energy constantly as there would be no Earth shadowing of the PowerSats. This is possible because the ablatness of the Earth at the equator will cause the PowerSats to precess at one degree per day and therefore maintain a position which matches the Earth's rotation around the Sun. These PowerSats will beam microwave energy to Space Tugs in equatorial or other orbits.

This ability to beam energy to the Space Tugs allows their mass to be very low because the energy production system would not be located on the Space Tugs. This reduction in mass would be a factor of approximately 30 times less than Solar, Chemical or Nuclear vehicles.

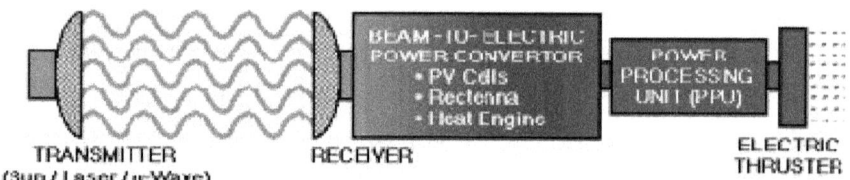

Beamed Electric Propulsion System

Electric propulsion (EP) provides much lower thrust levels than conventional chemical propulsion (CP) does, but much higher specific impulse. This means that an electric propulsion (EP) device must thrust for a longer period to produce a desired change in trajectory or velocity; however, the higher specific impulse enables a spacecraft using EP to carry out a mission with relatively little propellant and, in the case of a deep-space probe, to build up a high final velocity.

The key is providing the space tug will lots of energy. By beaming the energy to the space tug (Figure 1) the space tug doesn't need to carry the energy production system with it. This will make it much lower mass and therefore much faster than other concepts - if you can provide it with enough energy. At greater scale, i.e., multi-megawatt or even gigawatt level, WPT space tugs can outperform even chemical rockets in thrust.

Electric propulsion has become a cost effective and sound engineering solution for many space applications. Two of the main reasons why are its increased commercial availability and the opportunity it affords to perform the same task as conventional chemical propulsion systems while reducing the portion of the spacecraft's mass required for that task.

Electric propulsion systems have been tested on ground and in space since the 1960s and a wide variety of system types are available or have flown. Beamed microwave power represents a technological breakthrough because the mass of the rectenna on board the space vehicle is about equal to the mass of the electric thrusters, as contrasted to about thirty times as much for nuclear or photovoltaic. The envisioned hybrid WPT/Solar space tug uses both beamed energy and direct drive solar to move payloads beyond low Earth orbit (LEO).

Destinations could be GEO, Earth orbiting Colonies, Moon, Mars, etc. With 0.75 MW of power payloads of up to 60,000kg could be moved to lunar orbit. Unlike solar only space tugs which have the problem of Earth shadowing, which limits the amount of time the Ion drives can operate, the hybrid space tug can operate constantly due to the beamed energy it receives from the PowerSat.

This allows the Space Tug to accelerate much faster. While the power will vary depending on solar input, the tugs can always operate at a minimum power level of 0.5MW until it reaches the distance limit of the microwave beam. At that point the tug would be solar only at 0.25MW. Since the space tug receives most of its energy from the PowerSat, it would use a rectenna to capture the microwave energy.

Rectenna can be very low mass and operate with very good efficiency. The efficiency of the rectenna is estimated at a minimum of 80% conversion efficiency. Removing the mass of the solar arrays from the tug makes it a very low mass vehicle. Because it is low mass it can accelerate much faster than solar powered tugs which have limited acceleration due to Earth shadowing.

Additionally, since the energy is beamed to the tug constantly the ion drives can operate continuously. Most of the time the tug could operate at 0.75MW and this ability to constantly thrust combined with the low mass of the space tug will substantially reduce trip times to GEO, Moon and Mars.

Gridded electrostatic ion thruster research (past/present):

• NASA Solar electric propulsion Technology Application Readiness (NSTAR)

• NASA's Evolutionary Xenon Thruster (NEXT)

• Nuclear Electric Xenon Ion System (NEXIS)

• High Power Electric Propulsion (HiPEP)

• EADS Radio-Frequency Ion Thruster (RIT)

• Dual-Stage 4-Grid (DS4G)

Electric thrusters tend to produce low thrust which results in low acceleration. Using 1 g is 9.81 m/s/s; $F = m a$ or $a = F/m$

An NSTAR thruster producing a thrust (=force) of 92 mN will accelerate a satellite with a mass of 1000 kg by 0.092 / 1000 = 0.000092 m/s/s (or 9.38E-6 g).

Gridded electrostatic ion thrusters commonly utilize xenon gas. This gas has no charge and is ionized by bombarding it with energetic electrons. These electrons can be provided from a hot cathode filament and accelerated in the electrical field of the cathode fall to the anode (Kaufman type ion thruster).

Alternatively, the electrons can be accelerated by the oscillating electric field induced by an alternating magnetic field of a coil, which results in a self-sustaining discharge and omits any cathode (radiofrequency ion thruster). The positively charged ions are extracted by an extraction system consisting of 2 or 3 multi-aperture grids.

After entering the grid system via the plasma sheath the ions are accelerated due to the potential difference between the first and second grid (named screen and accelerator grid) to the final ion energy of typically 1-2 keV, thereby generating the thrust.

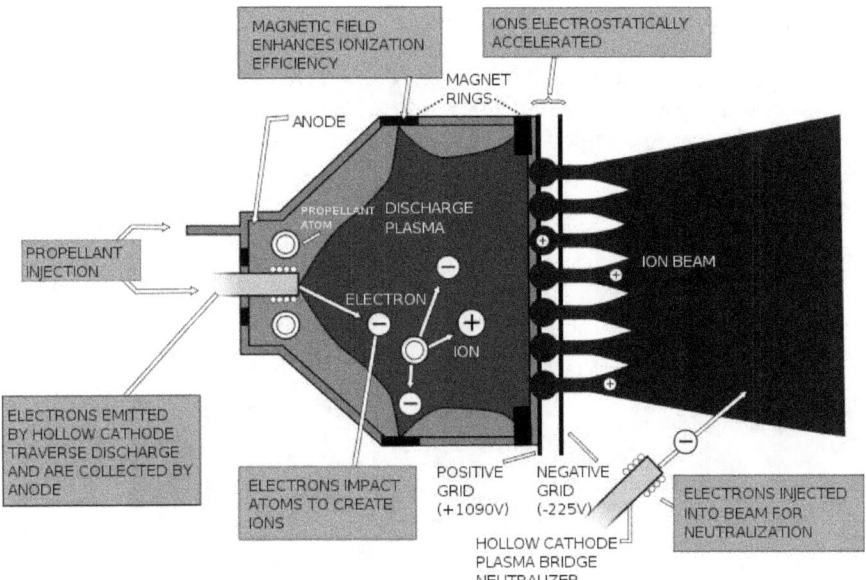

Ion thrusters are frequently quoted with efficiency metric. This efficiency is the kinetic energy of the exhaust jet emitted per second divided by the electrical power into the device.

The actual overall system energy efficiency in use is determined by the propulsive efficiency, which depends on vehicle speed and exhaust speed. Some thrusters can vary exhaust speed in operation, but all can be designed with different exhaust speeds. At the lower end of Isps the overall efficiency drops because the ionization takes up larger percentage energy, and at the high end propulsive efficiency is reduced. Optimal efficiencies and exhaust velocities can thus be calculated for any given mission to give minimum overall cost.

Inter-orbital Vehicle-Assumptions and specifications[24]

1. Makeup of the mass of the empty vehicle
   - Rectenna — 14 000 kg
   - Ion engines — 14 000 kg
   - Structure, power conditioning and propellant tanks
   - Total mass — 42 000 kg
2. Propulsion specifications
   - Rectenna dc power output — 20 000 kW
   - Rectenna dc power density — 400 W/m2
   - Rectenna area — 50 000 m2
   - Ion thruster
     - Propellant — Xenon
     - Specific Impulse — 4200s
     - Physical size — 50 cm diameter
     - Beam power — 30 kW each
     - No. of thrusters — 500
     - Mass of each thruster — 28 kg

   Total propulsive force — 750 N

   Vehicle acceleration (empty) — 0.0178 m/s^2

Mass breakdown for a 2MW beamed energy space tug

Hybrid WPT/Solar Space Tug:

Cargo Satellites (CargoSats) are cargo carriers similar to train boxcars. They are loaded at factories or supply depots in space. The Cargo Satellites have small chemical or ion. They can be transported by a wide variety of transport systems including manned and unmanned vehicles and moved around as necessary. CargoSats launched into space can remain in position for months and even years.

---

[24] Brown W. (1992). Beamed Microwave Power Transmission and its Application to Space, IEEE Transactions on Microwave theory and thechniques, Vol40. No6, 1239-1249

This would make space exploration simpler, less costly and safer since a supply of food, water, oxygen, propellant, etc can always be close by. CargoSats would be cheaper than other means of cargo transport because they are basically just storage containers. They would also be highly reusable.

By removing most of the cargo, habitat space and lunar roving vehicles, etc from the manned vehicle we can make the manned system much smaller than currently planned. Since the Orion Capsule and Service Stage use hypergolic propellants, they can be launched without the crew and maintained in orbit for an extend period of time. We can also make the cargo vehicle small since it won't need the Orion capsule and Service Module. The cargo version can take a slower 100 day route to the moon and save some propellant. This would also save considerable mass.

Launching the supplies first would also be safer than the current plans since plenty of cargo can be put on the lunar surface before the astronauts head out. In the current plan we launch one massive transport vehicle toward the moon and if something goes wrong, well just remember Apollo 13 and hope nothing goes wrong. Personally, I wouldn't want to get stuck on the Moon with no food or water and no way home.

Every ton of propellant that does not have to be on the spacecraft can be replaced by an equivalent ton of useful spacecraft structure. It would allow post-launch checkout of all spacecraft systems before departure from vicinity of Earth. It would require temporary docking to load fuel, but no assembly of modules in space. The key to this tactic is the orbital fuel (or propellant) depot. The Spaceship could be returned to earth and refueled in orbit using a Space Fuel Deport. A large portion of every spacecraft is propellant, and there is no need to launch fully-fueled vehicles if the fuel could be already waiting for them in orbit.

A depot could also be used at a transfer location, such as Lunar Orbit, or the Earth-Moon Lagrange L1 or L2 points. A depot would even be needed in Mars orbit once extensive human exploration operations start. Because most of the mass necessary to get to the moon is propellant this kind of space infrastructure would eliminate the need for a heavy-lift launcher altogether, increasing the launch rate of smaller, cheaper vehicles, which in turn could cut costs for getting to the moon and, eventually, Mars.

Space tugs are multi-use spacecraft that attach to and propel other spacecraft or payloads, modifying their orbits. Tugs have the potential to support a wide variety of space mission types, such as retiring geostationary communications satellites, moving cargo to space stations, Moon, Mars and other destinations, cleaning up space debris or on-orbit assembly tasks. A key feature of the concept is that the Tug is fully reusable and will deliver the cost benefits of reusability compared to expendable systems.

The Tug is launched just once and remains in orbit throughout its lifetime, where it is never subjected to the stress of repeated re-entry and re-launch. This eliminates the recurring cost of subsystem checkout, refurbishment, and recertification, required for concepts that must repeatedly survive the stress of launch and reentry. This paper will discuss a Beamed Energy In-Space Transportation System or Space Tug using small orbiting power satellites (PowerSats) located in a unique orbit and beaming microwave energy to the orbiting space tug.

These space tugs will move payloads throughout the inner solar system in support of human settlement. The PowerSats will be located in a circular sun-synchronous orbit (SS-O) at 2,722 kilometers (km) at 110 degrees to the equator. These small PowerSats will generate a 0.5 MW microwave beam and transmit the energy to hybrid space tugs. Due to the PowerSat location in the sun-sync orbit two of them can provide constant power to up to two space tugs at the same time.

Adding more PowerSats will decrease beam distances while powering even more space tugs. The space tugs are hybrid because they have both a wireless power transmission (WPT) rectenna and 0.25 MW of on-board direct drive solar cells. The combined energy will power a 0.75 MW Ion Drive Propulsion system on the space tug.

A beamed energy in-space transportation system is a low cost way to undertake massive space settlement in a way that is logical, technically possible and potentially very profitable. While there have been many past concepts for space tugs, sometimes called orbital transfer vehicles (OTV) most are now outdated concepts based on chemical propellants. Such systems are now outdated because electric propulsion (EP) is far superior.

The propellant needs for chemical tugs are 10-15 times greater than for solar electric tugs and beamed energy tugs can be 20-30 times lower mass than solar electric tugs (1). Therefore, beamed energy systems using microwaves are potentially a much lower mass solution for in-space transportation when large amounts of cargo need to be transported, such as supporting bases on the Moon or Mars. A quick look at the up-mass requirements in supporting the International Space Station (ISS) will give you a good idea how challenging supporting bases on the Moon or Mars will be.

Beamed energy concepts offer an alternative for an advanced low mass propulsion system. The use of a remote power source reduces the weight of the propulsion system in flight and this, combined with the high performance of electric propulsion, provides for increased payload delivery. There are different types of beamed energy propulsion concepts. This paper discusses a microwave beam generated by Power Satellites (PowerSats) located in a circular sun-synchronous orbit (SS-O) at 2,722 kilometers (km) at 110 degrees around the Earth. By locating the PowerSats in this orbit they can produce energy constantly as there would be no Earth shadowing of the PowerSats.

This is possible because the oblateness of the Earth at the equator will cause the PowerSats to precess at one degree per day and therefore maintain a position which matches the Earth's rotation around the Sun. These PowerSats will beam microwave energy to Space Tugs in equatorial or other orbits. This ability to beam energy to the Space Tugs allows their mass to be very low because the energy production system would not be located on the Space Tugs. This reduction in mass would be a factor of approximately 20-30 times less than Solar, Chemical or Nuclear vehicles (1).

In 1986, Graeme Aston of NASA's Jet Propulsion Laboratory proposed a lunar transportation system based on solar-electric propulsion (SEP) space tugs for ferrying moon base elements and cargo between Earth and lunar orbit. (4) Each cargo module would be delivered to low lunar orbit by the SEP tug and then land on the lunar surface using a small built-in descent rocket. After the base had been completed, the fleet of SEP ferry vehicles would continue transporting supplies, crew consumables and equipment. Four vehicles operating simultaneously could deliver a 20,000-kilogram (kg) payloads to low lunar orbit every 100 days. The reusable 7,700 kg (dry mass) ion-propelled lunar ferry would deliver a 20 metric ton payload to a 100 km lunar orbit and return under its own power to a 300 km low Earth orbit. Only 7.3 metric tons of xenon propellant would be required per flight.

The drawback was the total flight time of one year (the system would be used primarily for unmanned logistics support) but the payload fraction would be very high -- 60 %. This was because of the extremely high specific impulse (4,500 s) of the engines, reducing the required spacecraft mass in low Earth orbit (and hence launch costs) by 50% compared with a Shuttle-derived HLLV. Power for ten engines of 1 N thrust each was generated by two giant 12 x 61 meter solar arrays each producing 150 KW of power.

Aston claimed solar electric propulsion was more economical than nuclear power as long as the required power output was less than about 1000 KW. (4) Advances in solar cell technology since his proposal would push this out to perhaps 10,000 KW. Therefore, there is no reason to even consider nuclear space tugs.

Solar Electric Space Tug:

• Lunar Ferry Vehicle Initial Mass (includes Module and all Propellant) 35 000 Kg

• Moon base Cargo Module 20 000 Kg

• Xenon Propellant consumed by ion engines 7 300 kg

• S.E.P. Lunar Ferry Dry Mass (Less Cargo) 7 700 kg

• Low Earth Orbit (Leo) Parking Altitude 500 km

• Low Lunar Orbit (Llo) Parking Altitude 100 km

• Total Trip Time (Leo-Llo-Leo) 370 days

• Solar Array Power To Ion Engines 300 kW

• Solar Array Power Density 0.2 kW/m2

• Solar Array Specific Mass (Beginning Of Life - End Of Life) 5 To 6 kg / kW

• Total Ion System Efficiency 0.75

• Ion Engine Specific Impulse 4500s

• Ion Engine Specific Mass 6 kg / kW

• Number of Ion Engines (including spares) 12

In reference 2 we see a solar powered space tug operating at 0.7 MW to move 60,000kg of cargo to the Moon and based on the technology available by 2010. The trip time to the Moon is estimated at almost one year. The shadow problem for solar space tugs limits initial acceleration of the vehicle away from the Earth's gravity well. When close to the earth the shadow time would be about 40% per orbit. Even at 10,000km the shadow time would be about 12% per orbit. This limits the amount of time that electric thrusters can operate and is the main reason why trip times to the GEO, L1, etc. using solar powered space tugs would so long.

Mission Condition	Mass, kg	ΔV, m/sec
Initial Mass in LEO (IMLEO)	100,130	-
Mass at Lunar Insertion	78,847	5155
Mass at Lunar Capture	78,524	89
Mass in Intermediate LO	76,473	556
Mass at 400 Km LLO	73,270	707
Payload Mass to LLO	61,752	-
SETV Mass leaving LLO	11,518	-
SETV Inert Mass back in LEO	9,757	6507

Lunar SETV Performance table (Ave Isp = 2200 sec outward & 4000 sec return)

The Solar Electric Space (SEP) Tug uses a space docking system, solar arrays and ion drives so the stage can remain in orbit for extended periods of times (years) and transport cargo to the International Space Station (ISS) and other space platforms.

This first Tug concept uses a modified version of the Alenia Spazio mini-MPLM to transport pressurized cargo. Unlike current cargo transport modules (Progress, HTV, ATV) these Cargo Modules (VCM) are designed to remain is space for extended periods (years) and can be berthed or docked to the ISS depending on need.

The Tug can remain in orbit as a temporary or fixed element of a space station or house in-space manufacturing systems. The Space Tug transport system has a greater cargo capability then the Progress transport vehicle and is the most cost effective solution for ISS and other space platform servicing when combined with launch vehicles (LVs) delivering cargo in Leo Earth Orbit (LEO). The reusable Space Tug does not compete with NASA's Exploration Hardware but supports the total vision by offering a flexible and cost effective solution for up mass delivery.

This allows for the pre-positioning of cargo and systems in space or on the lunar surface, which in turn allows for longer and more flexible missions. This also reduces the risk of potential loss of crew on the surface as supplies and support systems can provide for extended mission times if necessary.

While this proposed Space Tug uses much existing components and systems and existing launch vehicles, its operations and uses are unique. It is based on standardization of system components for multiple missions and a wide variety of mission types, such as Station servicing and lunar/mars cargo transport. This includes the ability to transport cargo from low Earth orbit (LEO) to higher orbits in multiple trips and satellite, long-term in-space storage of cargo, and use of the Cargo Modules as permanent components of space stations or other platforms including in-space manufacturing facilities.

The Space Tug can make as many as ten trips between LEO and the Space Station before needing refueling. This ability for reusable space tug operations substantially lowers cost verse the Progress, ATV and HTV alone by increasing the total mass lifted into space.

The Space Tug is launched with a small cargo module on the first mission and flies directly to the Station; subsequent missions (2-9) launch a larger cargo module into a 250km orbit using a government or commercial launch vehicle. The Tug maneuvers to the lower orbit, docks with these larger modules and transports them to the Space Station.

The use of ion drives allows for long term loitering in orbit while waiting for the next mission. It also allows large orbit adjustments at minimum propellant usage. Ion drives can reduce mission propellant mass by 90 percent compared to conventional liquid or solid propellant vehicles.The modular architecture allows the tugs to be staged. Staging would allow the transfer of much greater cargo mass and/or allow missions to be staged for lunar/mars or other deep space destinations. Another advantage of staging is that it recovers most of the components much earlier as you don't have to wait months or years for the return of the booster stages.

The Cargo Modules can be docked or berthed together to form components of larger space stations or for free flyers for in-space manufacturing in LEO, Lunar L1 or other locations. The Modules can have one to 6 docking ports depending on application. The docking ports can be active (for docking) or passive (for berthing) in any desired combination. A wide variety of assembly combinations is therefore possible. The external attach points allows for the integration of trusses or other exterior items such as fixed or movable solar arrays.

The Space Tug can also support many of the Exploration Architectural Elements including but not limited to: Food, Water, Basic Power, Habitation, Rovers, Robotic Missions, ISRU and of course General Logistics. The Food Modules and Venture Space Tug are proposed as private initiatives to support both commercial (such as Bigelow Aerospace) and NASA needs.

A tug might also make practical a single stage to orbit (SSTO) reusable launch vehicle (RLV). Since a first generation SSTO will most likely provide a very small payload capacity, it would help if it only had to reach a low orbit where it would transfer cargo/crew to a tug and also pick up cargo/crew to bring back from orbit. Even with small payloads, the simplicity of SSTO RLV operations might lead to reduced LEO delivery costs when combined with a tug.

Regardless of vehicle size, you'll get a lot more payloads to the station if you use the tug (some of the extra payload is used to refuel the tug). The hybrid WPT/Solar space tug uses both beamed energy and direct drive solar to move payloads beyond low Earth orbit (LEO). Destinations could be GEO, Earth orbiting Colonies, Moon, Mars, etc. The PowerSat uses solar arrays to collect the energy from the sun. Since the PowerSat is located in a sun-sync orbit it can generate power constantly because the PowerSats never enters the Earth's shadow.

With 0.75 MW of power payloads of up to 60,000kg could be moved to lunar orbit. Unlike solar only space tugs which have the problem of Earth shadowing, which limits the amount of time the Ion drives can operate, the hybrid space tug can operate constantly due to the beamed energy it receives from the PowerSat. While the power will vary depending on solar input, the tugs can always operate at a minimum power level of 0.5MW until it reaches the distance limit of the microwave beam.

At that point the tug would be solar only at By removing most of the solar arrays the tug becomes very low mass. Because it is low mass it can accelerate much faster than solar powered tugs. Additionally, since the energy is beamed to the tug constantly the ion drives can operate continuously. By removing most of the solar arrays the tug becomes very low mass.0.25MW. Most of the time the tug could operate at 0.75MW.

Since the space tug receives most of its energy from the PowerSat, it would use a rectenna to capture the microwave energy. Rectenna can be very low mass and operate with very good efficiency. The efficiency of the rectenna is estimated at a minimum of 80% conversion efficiency.

To generate a 0.5 MW microwave beam the PowerSat would need to collect about 1 MW of solar energy for an energy conversion efficiency of 50%. There are a number of ways to generate energy from the sun including flat panels, concentrated panels, thin film and solar thermal. The microwave beam could be generated by small discrete microwave components or by a large constant output single tube Klystron. Since Klystrons are generally designed to operate in pulse mode two pulsing Klystrons could be used with equally timed pulses to generate a constant beam.

The PowerSats mass would be low enough that it would be lunched on a single launch vehicle. There are a number of existing launch vehicles that could launch the PowerSat. These include the Atlas, Delta, Araine, Proton, H-IIB and Falcon 9.

The PowerSat would generate a microwave beam at 40 Ghz. The microwave energy would be beamed to the space tug. There are no power loses in space, however, the greater the distance the greater the beam spread so there is a limit to how far the beam can travel and still provide energy to the space tug. Since the PowerSat is located in an obit at 2,722km it would beam the microwaves to the Tug in LEO.

As the tug moves further out and passes the PowerSat the PowerSat can continue beaming out to a distance of some 12,000 m. Therefore, the total distance would be about 14,722km under which the tug could be powered by the beamed energy. Past this point the space tug would operate in solar only mode. On the return trip the tug could start receiving energy from the PowerSat and use the energy to slow down.

PowerSat

No. PowerSats	2-4	
Location	2,722 km	
Solar Concentration (SLA)	12	suns
Solar Conversion Efficiency	32	%
Transmitter Efficiency	80	%
Transmitter Frequency	40	GHz
Beam Distance	15,000	km

While space solar power (SSP) requires very large PowerSats, beamed power in space can use very small PowerSats. A beamed energy in-space transportation System could be considered a precursor and even a technology demonstrator for SSP. The technology is basically the same for both and it is simply a matter of scale and beam frequency. Therefore, developing a beamed energy in-space transportation System could led to faster implementation of SSP and a solution to the Earth's energy problems.

Low cost in-space transportation is a critical requirement to all space projects beyond LEO. At the multi-megawatt level beamed energy has the potential to transport millions of humans and massive amounts cargo rapidly throughout the inner solar system.

This is the technology of the future and it is available now if we want it. It is possible to start at PowerSat levels less than 1 MW and then deploy larger and more capable systems. Spaced throughout the inner solar system and powering a space highway to Free Space Colonies, the Moon and Mars these PowerSats and WPT Space Tugs would provide a very low cost transportation system.

With the ability to move though space at low cost the demand for launch services will increase dramatically. Low cost in-space transportation is an enabling technology for the development of low cost Earth to LEO launch systems. Certainly the technology to deploy solar electric space tugs already exists and has for some time. The technology for beam powered space tugs is also available. Neither has been deployed to date. This lack of adequate in-space transportation severely limits our ability to settle or even explore the near Earth solar system.

The development of electric in-space transportation has the potential to increase our ability to explore deep into space and open space to massive human settlement. Commercially developed in-space transportation systems can be deployed using existing technology. A beamed energy in-space transportation system is the lowest mass solution because it can power multiple space vehicles with continuous energy. Such a system could be deployed at small scale and low cost. The development of beamed energy space tugs is a prerequisite to any space settlement effort.[25]

Power generation is one of the crucial elements of space vehicles and of future infrastructures on planets and moons. The increased demand for power faces many constraints, in particular the sizing of the power generation system.

---

[25] Jones, R. (2012), Beamed Energy In-Space Transportation System for Near Space Colonization, earthspaceagency.org

The SPS Space Grid system is a candidate solution to deliver power to space vehicles or to elements on planetary surfaces. Beaming energy to spacecraft could lower spacecraft mass and improve mission-economic potential. A BSEP system with SS-O PowerSats beaming to equatorial Ion Spaceship would reduce the ship's mass by a factor of 30 compared to direct drive solar or nuclear ships. This promises a significant reduction in the cost of space transportation.

Reusable in-space transportation systems must be capable of both high fuel efficiency and high utilization of capacity, or economic costs will remain unacceptably high. BSEP systems can provide high fuel efficiency and with enough high thrust to support cargo and crewed missions. The major contribution of beamed power to the development of space is its unique ability to transfer energy across long distances and across large differences in gravitational potential. This technology can also be applied to a Spaceship in LEO orbit propelled by electric thrusters whose power is supplied by a microwave beam originating at SS-O. The development of a Space Grid using SS-O SBSP PowerSats with Equatorial orbiting ReflectorSats appears to offer all of the advantages of GEO PowerSats, including constant power production for base-load energy.

Additionally, it appears to offer many advantages over GEO stationing including, no LEO-GEO transportation, ease of maintenance due to closer positioning, much less mass to orbit due to transmitter sizing and the ability to direct more energy to the day side of the Earth. While operationally the concept is more complex than GEO stationing, this is overshadowed by the large reduction in the minimum size verse GEO since the minimum size of the PowerSat is reduced from 5GW to only 2GW. What this means is that you can start producing power in space by launching only 11,494,720 kg rather than 34,000,000kg and this can move humanity closer to clean energy from space.

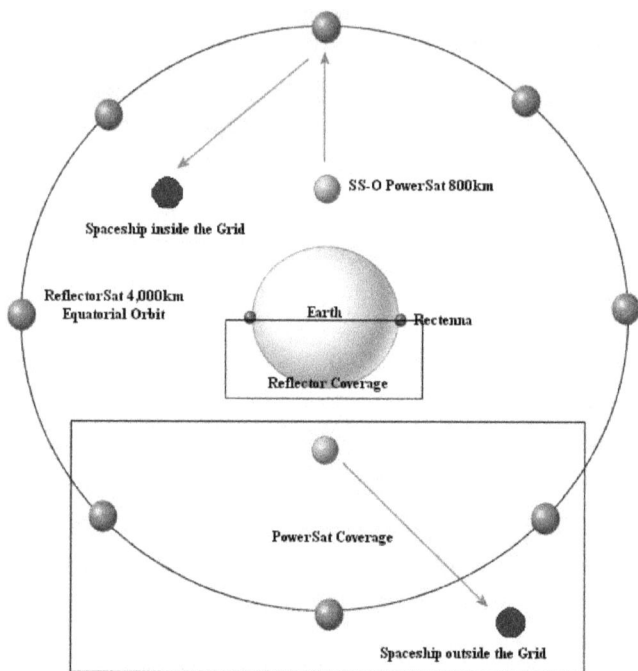

Looking down at the Earth's North Pole. The space grid is powering two orbiting spaceships, it shows one spaceship operating inside the grid and another spaceship operating outside the grid.

It also means that we can power in inner solar system and open space settlement to humanity. Let's simplify the argument and assume that regardless of location all systems would use the best available system for power production. Therefore, regardless of location the power system will be the same for all concepts. This then becomes an argument related to distance and therefore transmitter size and mass or does it?

The past assumption was that GEO was the best location was based on

1) Availability of power and

2) The ability to beam to a single location 24/7.

As has already been shown, there is at least on other location other than GEO that can provide power 24/7 power and that there are at least two ways to use this location. This location is a sun-synchronous dusk to dawn orbit. This obit can be 800km, 1,500km or even 2,722km and is therefore very close to the Earth. This orbit can be used with equatorial or Molniya based reflectors to deliver 24/7 power.

The argument against LEO and MEO PowerSats has mostly to do with power losses due to Earth shadowing which would require larger satellites or more satellites to make up for the lost shadow time. The assumption has been that this would result in a total mass greater than that of a GEO PowerSat. Is this argument valid? Actually, no it is not. While it would add mass in some cases, the mass would still be lower than that of a GEO PowerSat in most all cases. Therefore, the argument is false.

The idea that GEO is the best location based on mass to orbit has been debunked. Now let's go a little further. The other cost of concern is delivery to orbit. Typical communications satellite solar panels have a mass per kW of about 20 kg, so with current launch costs of $10,000/kg that comes to $200/Watt, or a hundred times too large to be competitive at the utility level. Bringing that number down requires improvements in mass per kW and cheaper access to space. Mass per kW is sensitive to solar power system design (3). The NASA/DOE reference design came to 10 kg/kW; more recent studies of light-weight design options have suggested mass could be as low as 1 kg/kW (4).

Competition in the commercial launch market already has some providers such as Sea Launch offering $4000-$5000 per kg prices to low earth orbit. Use of solar electric propulsion allows higher orbits at only slightly higher cost. Given the multi-trillion-dollar potential market for space-based power, increased funding for launch systems development to accelerate these improvements would also be a worthy investment.

We already have an immense fusion reactor working for us in our solar system, ultimately responsible for almost all our energy choices. All we really need to do is make better use of it by tapping into it more directly.

Typical reference designs involved a satellite in geostationary orbit, several kilometers on a side that used photovoltaic arrays to capture the sunlight, and then convert it into radio frequencies of 2.45 or 5.8 GHz where atmospheric transmission is very high, that were then beamed toward a reference signal on the Earth at intensities approximately 1/6th of noon sunlight. The beam was then received by a rectifying antenna and converted into electricity for the grid, delivering 5-10 gigawatts of electric power.

One of the keys to success of any MPT platform is the rectenna since it must rectify power as efficiently as possible. Work performed by James McSpadden at Teas A & M University in conjunction with JPL has led to a 2.45-GHz rectenna element designed for more than 85-percent RF-to-DC power conversion efficiency but that could, in theory, provide considerably higher efficiency if optimized to operate while oscillating at a higher frequency, such as the 3.3 GHz suggested by the researchers. The rectenna consists of a half-wave dipole antenna, two-section input lowpass filter, GaAs IMPATT diode, and output capacitor for shorting the RF power and using the diode.

The development of an economically viable space-based solar power (SBSP) system is critical to the Earth's future and for future space development. PowerSat technology is also critical to supporting sustainable private and government space ventures, including space lift, space exploration and space infrastructure development.

Such a system would greatly expand the need for space lift capability from small reusable launch vehicles for SBSP satellite maintenance to large expendable launch vehicles for deploying GW class SBSP satellites into orbit.

The technology needed for SBSP is also needed for in-space solar electric transportation systems needed for space colonization as the technology is the same. The hope has been that gradual improvement in Photovoltaic or other technologies such as thermal systems would solve the mass to orbit problem for SBSP systems. However, this in itself does not appear sufficient to make SBSP economically viable.

A new architectural option for SBSP uses a Sun -synchronous orbit (SS-O), wireless power transmission (WPT) and a space power relay (SPR). This new concept is called The Space Grid. The Space Grid relies on the use of two separate satellite constellations.

The power satellite (PowerSat) constellation is placed in SS-O dusk to dawn orbit at 800km and has access to constant sunlight and is used to produce the power. The Equatorial reflector satellite (ReflectorSat) constellation is in a 4,000km equatorial orbit and is used to distribute the power to the rectenna on the Earth's surface. The power is produced by the PowerSats in SS-O and beamed to the ReflectorSats in equatorial orbit and then bounced to the rectenna on the ground. This combination allows for the production and distribution of power to the Earth's surface without the problems normally associated with non-Geostationary (GEO) PowerSat concepts and without having to place the PowerSats in GEO.

The Space Grid reduces the mass of a PowerSat transmitter by approximately 67% by moving it closer then past concepts GEO concepts and allows for higher power levels and therefore much smaller (60%) and less costly rectenna on the ground and reduces the minimum size from 5GW to only 2GW allowing quicker deployment of space energy to solve the Earth's energy problems. WPT transmission could be microwave or laser but for this paper microwave will be used for easier comparison with past concepts.

Chapter Ten

Large Scale Space Settlement

Beamed energy concepts offer an alternative for an advanced propulsion system. The use of a remote power source reduces the weight of the propulsion system in flight and this, combined with the high performance, provides significant payload gains. Going to the Moon, Mars and the stars requires a vision for the future and innovative technology development to take us there.

The high cost of space transportation coupled with unreliability is a handicap to development of the final frontier. Imagine the possibilities when space transportation becomes safe and affordable for everyone. Whether it's living and working in space, exploring new worlds or just leaving the planet for a vacation, the opportunities for business and pleasure on the space frontier are endless.

---

[26] Image Retrieved from http://www.mc2quantum.com/wp-content/gallery/space-stations/international-space-station-iss-04.jpg

Electric Space: Space-based Solar Power Technologies & Applications

Our dreams of everyday life in space and its promise for a better life on Earth are hostage to the high cost of space transportation. Dramatic improvements are required to make space transportation safer and more affordable.

Future space launch vehicles must be safer, more reliable, simpler and highly reusable. This applies not only to ETO vehicles but also to vehicle operating is space.

With the ability to move through space at low cost the demand for launch services will increase dramatically. Low cost in-space transportation is an enabling technology for the development of low cost Earth to LEO launch systems. We will finally be able to build and support bases on the Moon and also Mars. Asteroid mining would be possible.

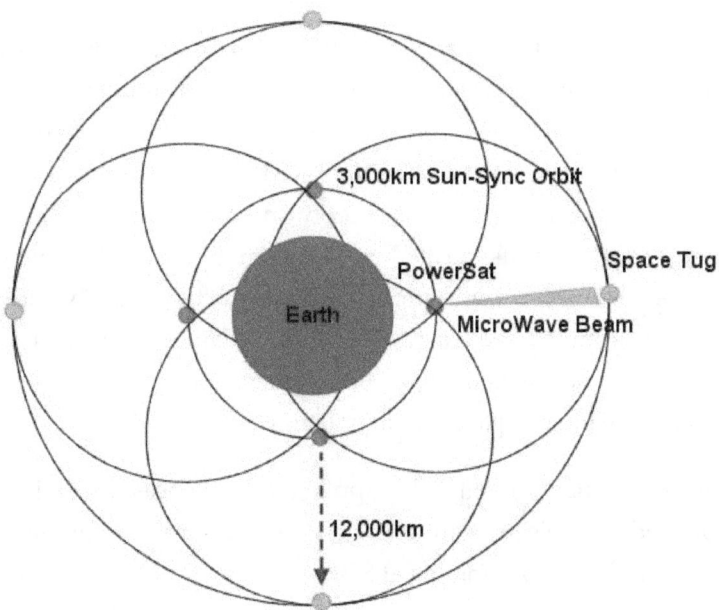

Electric Highway using four PowerSats to power hybrid WPT/Electric Space Tugs. In this case the microwave beam distance is 12,000km, past this point the Space Tugs would operate in solar only mode. The four larger inner circles show the beam range of each PowerSat.

While space solar power (SSP) for Earth energy requires very large PowerSats, beamed power in space can use very small PowerSats. A beamed energy in-space transportation System could be considered a precursor and even a technology demonstrator for SSP.

The technology is basically the same for both and it is simply a matter of scale and beam frequency. Therefore, developing a beamed energy in-space transportation system could lead to faster implementation of SSP and a solution to the Earth's energy problems.

Certainly the technology to deploy solar electric space tugs already exists and has for some time. The technology for beamed energy space tugs is also available. Neither has been deployed to date. This lack of adequate in-space transportation severely limits our ability to settle or even explore the near Earth solar system.

The development of beamed electric in-space transportation has the potential to increase our ability to explore deep into space and open space to massive human settlement. Commercially developed in-space transportation systems can be deployed using existing or very near term technology. A beamed energy in-space transportation system based on PowerSats located at 2,722km SS-O is the lowest-mass solution because it can supply multiple electric space vehicles with continuous energy. Such a system could be deployed at small scale and low cost. The development of beamed energy space tugs is a prerequisite to any space settlement effort.

This technology can also be applied to a Spaceship in LEO orbit propelled by electric thrusters whose power is supplied by a microwave beam originating at SS-O.

Electric Space: Space-based Solar Power Technologies & Applications

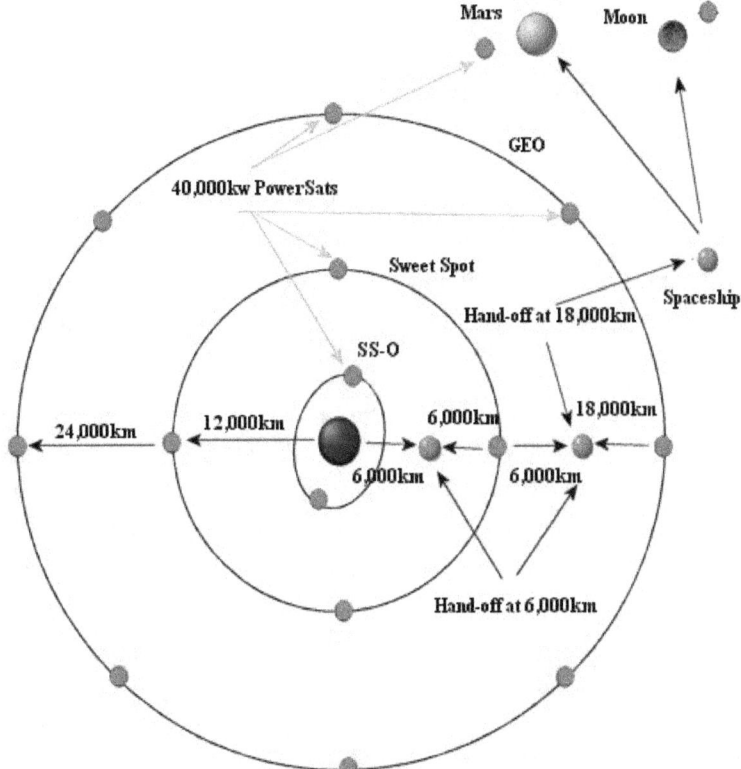

Electric Superhighway in space

Space tugs are multi-use spacecraft that attach to and propel other spacecraft or payloads, modifying their orbits. The PowerSats will be located in a circular sun-synchronous orbit (SS-O) at 2,722 kilometers (km) at 112 degrees to the equator.

These small PowerSats will generate a 0.5 MW microwave beam and transmit the energy to hybrid space tugs. A beamed energy in-space transportation system is a low cost way to undertake massive space settlement in a way that is logical, technically possible and potentially very profitable.

Rectenna in Space:

Using a rectenna for microwave beam reception offers a potentially very low mass solution to in-space transportation. Using a rectenna to generate the energy for the space vehicle would reduce the vehicle's mass by a factor of 30 compared to chemical, solar or nuclear powered vehicles. It would have similar mass reductions compared to laser based beamed energy systems because using laser frequency band-gapped cells at even 50% efficiency would be little better than concentrated solar band-gapped cells at the same efficiency rating. The only mass savings would come come from not needing a light reflector and those are potentially very low mass structures. Rectenna efficiency can potentially exceed 80% which is 30% better at hugely lower mass than a laser based system.

Powering Spaceships:

Much of the technology for Space Base Solar Power (SBSP) and Solar Electric Propulsion (SEP) are the same. This is especially true when considering beamed solar Electric propulsion (BSEP). We will look at a new concepts for SBSP but will also show how the same concepts can be used for space settlement, i.e., space colonization by supplying massive amounts of energy to orbiting spaceships.

Wireless Power Transmission (WPT) and it relation to space may be thought of as extending our two dimensional power transmission networks on the Earth to space and to other planets and space vehicles.( ) Such a system could be used for a wide variety of applications. One such application would be providing large amounts of power for an electric spaceship needed for an in-space transportation system. Electric propulsion has long been recognized for its benefits if there were a suitable energy source for the large amounts of power required by electric thrusters.

Conventional prime power sources in space are massive relative to electric thrusters and must be accelerated along with the less massive parts of the vehicle. Further, they are expensive and costly to transport into space. In contrast, beamed microwave power removes the prime power source from the vehicle and therefore has a very low mass relative to other potential prime power sources in space, including chemical, nuclear and solar electric.

The combination of WPT and electric thruster technology would make it possible to replace conventional chemical rocket propulsion for missions beyond low Earth orbit (LEO) with enormous economic and safety benefits. It is interesting to note that the technology required for WPT SBSP is the same as that required for WPT BSEP systems. By pursuing SBSP to supply the Earth with energy we are also developing the technology for large scale colonization of the solar system at the same time.

The primary difference is simply one of scale. PowerSats for Earth energy would be very large systems of the gigawatt scale; however, BSEP systems can be very small at less than a megawatt. This means that BSEP systems can be considered technology demonstrators for the much larger SBSP PowerSats. That being the case we can develop a stepping stone approach starting with small BSEP systems and progress toward larger SBSP systems.

Low cost in-space transportation is a critical requirement to all space projects beyond LEO. At the multi-megawatt level beamed energy has the potential to transport millions of humans and massive amounts cargo rapidly throughout the inner solar system. This is the technology of the future and it is available now if we want it. It is possible to start at PowerSat levels less than 1 MW and then deploy larger and more capable systems. Spaced throughout the inner solar system and powering a space highway to Free Space Colonies, the Moon and Mars these PowerSats and WPT Space Tugs would provide a very low cost transportation system.

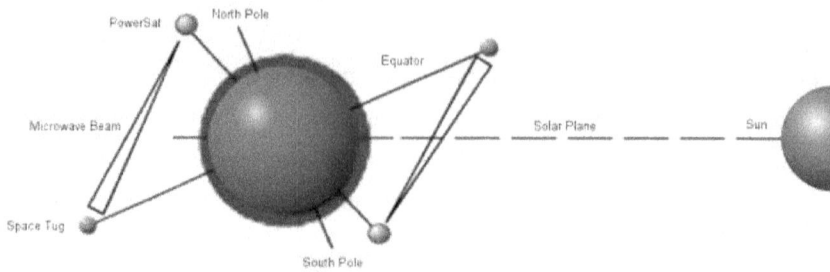

Two PowerSats powering two Space Tugs

A unique in-space transportation is required for initial system deployment, which takes place in LEO. An extremely affordable LEO-to-GEO in-space transportation system is vitally important to this concept. Power beaming from geostationary orbit by microwaves has the added difficulty that the required "optical aperture" sizes must be very large. The 1978 NASA SPS study required a 1km diameter transmitting antenna, and a 10 km diameter receiving rectenna, for a microwave beam at 2.45 GHz frequencies.

Mars Super Cheap:

Past proposals to travel to Mars, such as Mars Direct and others, have been based on the use of chemical propellants. Such proposals seem to be catered to NASA's past and present plans to construct a large heavy lift launch vehicle (HLLV). In truth, such a vehicle is not required to travel to Mars and the proposed development of a HLLV will be hugely expensive and is completely unnecessary.

A much more cost effective approach would be to use a hybrid WPT/SEP space tugs to transport cargo and crew to Mars. A hybrid WPT/SEP space tug combines wireless beamed energy with solar energy.

Using WPT/SEP space tugs creates reusable space architectures capable of multiple Mars missions. Since WPT/SEP space tugs are both more efficient then chemical rockets and less expensive than expendable space architectures based on chemical rockets they provide an inexpensive alternative to traveling to Mars. Such an approach would eliminate the need for an HLLV all together and free up some $20 billion by not building the proposed Senate Launch Vehicle (SLS). An architecture based on SEP space tugs would make going to Mars super cheap.

The scale of the required Earth-to-orbit (ETO) launch capability is determined by the mass of the largest payload intended for the Martian surface. The nominal design mass for individual packages to be landed on Mars in the NASA Reference Mission is 50 tones for a crew habitat sized for six people that is transferred on a high-energy orbit. This requires the capability for a single launch vehicle to be from about 200 to 225 tones to LEO. However, it the crew and most of the cargo were sent separately the size each mission could be substantially reduced.

The split mission strategy takes advantage of the currently available capability to successfully fly and land automated spacecraft on another planet. Such capability can be used to deliver supplies and equipment to support human missions without a crew being present. By using this capability to deliver cargo the size of the transportation system (both launch vehicles and upper stages) for any one mission becomes smaller and thus less expensive to develop and manufacture. In addition, these cargo missions can be sent on the absolute minimum energy trajectories between Earth and Moon/Mars because there is no time-critical or life support critical element on board for most cargo missions.

However, the total number of launches increases under this strategy which offsets at least part of the cost savings due to the increased number of transportation elements that must be used. However, if the components of the transportation system where reusable the cost would be substantially reduced over several missions and the cost of developing super large launch vehicles can be avoided.

By splitting the missions into cargo and crew flights, infrastructure can be set up and operated before committing a crew to a flight. Operating this infrastructure for an extended period prior to launching a crew also improves the confidence of using the Lunar/Mars surface as a safe haven for the crew. It should be noted that Robert Zubern's proposal for Mars Direct uses a split mission concept with two vehicles.

This technology has been developed at breadboard level and can be demonstrated on robotic missions. It provides significant benefits to the mission by reducing launch mass from Earth and increasing robustness of surface systems where caches of consumables and surface vehicle fuels can be maintained. As discussed in the previous section, any technology that can reduce the amount of mass (and propellant is the largest single item on such a list) can do much to reduce life cycle cost.

This is accomplished primarily by reducing the size and number of launches from Earth and by providing a dual purpose infrastructure that not only provides propellants for a return trip but also supports crew activities and helps reduce risk. Mars missions differ from Space Shuttle and lunar missions in that once the crew is committed to launch, orbit mechanics force the crew to remain away from Earth for approximately 2 to 3 years. This imposes on all of the systems the need for a higher degree of reliability and maintainability or for multiple independent means of providing life-critical functions (collectively referred to as robustness).

There has been a tendency to view the Martian surface as the most hostile location for a crew during a Mars mission. However, of the three environments that a crew will encounter—Earth, interplanetary space, and the Martian surface—interplanetary space offers the highest potential for debilitating effects on the crew. Practicality dictates a relatively small habitable space for the crew during transit. To do otherwise causes a corresponding increase in the size and cost of the systems, primarily launch vehicles and transfer stages, associated with the transportation system.

But to confine the crew to a small habitable space for an extended duration can lead to cabin fever. Zero g has known debilitating effects on the human body that must be addressed. Radiation from a constant background and the threat of solar flares require that protection be adequate for background sources and that a safe haven be provided for extreme events. All of these threats have engineering solutions that can make the extended stay in interplanetary space a viable prospect for the crew. But the solutions typically require increases in size, mass, and complexity of the vehicle and the transportation elements that are used to move it from planet to planet.

NASA found that generating electric power for terrestrial consumer use was not the only potential application for space solar power. Other uses have been postulated, including power transmission to other space vehicles, power generation for lunar and Martian exploration, power for commercial space development such as communications satellites, and as a source of additional power to enhance the capabilities of such on-orbit facilities as the International Space Station (Grey, 2000). Making some or all of these uses of space solar power a reality requires developing, fielding, and making effective use of a number of complex technologies within a constrained budget.(5)

For the DoD specifically, beamed energy from space in quantities greater than 5 MWe has the potential to be a disruptive game changer on the battlefield.

SBSP and its enabling wireless power transmission technology could facilitate extremely flexible "energy on demand" for combat units and installations across an entire theater, while significantly reducing dependence on vulnerable over-land fuel deliveries. SBSP could also enable entirely new force structures and capabilities such as ultra long-endurance airborne or terrestrial surveillance or combat systems to include the individual soldier himself.

More routinely, SBSP could provide the ability to deliver rapid and sustainable humanitarian energy to a disaster area or to a local population undergoing nation-building activities. SBSP could also facilitate base "islanding" such that each installation has the ability to operate independent of vulnerable ground-based energy delivery infrastructures. In addition to helping American and allied defense establishments remain relevant over the entire 21st Century through more secure supply lines, perhaps the greatest military benefit of SBSP is to lessen the chances of conflict due to energy scarcity by providing access to a strategically secure energy supply.

SSO PowerSat:

The PowerSat uses solar arrays to collect the energy from the sun. Since the PowerSat is located in a sun-sync orbit it can generate power constantly because the PowerSats never enter the Earth's shadow. To generate a 0.5 MW microwave beam the PowerSat would need to collect about 1 MW of solar energy for an energy conversion efficiency of 50%. There are a number of ways to generate energy from the sun including flat panels, concentrated panels, thin film and solar thermal. Alternatives are possible and should be considered by any developer.

One option that looks promising is the Stretched Lens Array (SLA) being developed by NASA which uses a solar concentration of 12 suns.

The microwave beam could be generated by small discrete microwave components or by larger constant output single tube Klystron. Since Klystrons are generally designed to operate in pulse mode two or more pulsing Klystrons could be used with equally timed pulses to generate a constant beam.

The PowerSat mass would be low enough that it could be launched on a single launch vehicle. There are a number of existing launch vehicles that could launch the PowerSat. These include the Atlas, Delta, Araine, Proton, H-IIB and Falcon 9 and the proposed Falcon Heavy. The PowerSat would generate a microwave beam at 40 GHz. The microwave energy would be beamed to the space tug. There are no power loses in space, however, the greater the distance the greater the beam spread so there is a limit to how far the beam can travel and still provide energy to the space tug. Since the PowerSat is located in an obit at 2,722km it would beam the microwaves to the Tug in LEO.

As the tug moves further out and passes the PowerSat the PowerSat can continue beaming out to a distance of some 6,000km. Therefore, the total distance would be about 8,722km under which the tug could be powered by the beamed energy. Past this point the space tug would operate in solar only mode. On the return trip the tug could start receiving energy from the PowerSat and use the energy to slow down.

PowerSat Orbit Details:

The PowerSats would be located in a 2,722km circular SS-O orbit at 110.1 degrees. This orbit is different than the orbit SS-O orbit used by the Canadian RadarSats. The orbit has greater altitude and is much less polar. This SS-O can provide constant power generation by the PowerSat because the PowerSat solar arrays will always face the sun and the PowerSat will not enter the Earth's shadow.

Because the PowerSat is at a greater altitude it would have a good view of the Space Tug which will allow the Space Tug to receive near constant energy from the PowerSats. For 24/7 power two PowerSats would be needed.

- Altitude 2,722 km SS-O
- Retrograde Orbit
- Angle to Equator 110.1 degrees
- Precession Rate 1 degree per day
- Orbit Revolution Time 143 min
- Orbits per day 10.07
- Satellite Orbit Velocity 6.62 km/second
- No. PowerSats 2-4
- Solar Concentration (SLA) 2,000 suns
- Solar Cell Conversion Efficiency 32 %
- Transmitter Efficiency 80 %
- Transmitter Frequency 40 GHz
- Beam Distance 12,000 km

Microwave Generator:

With four PowerSat available to power space vehicles the maximum beam distance would only be about 7,000km when inside the PowerSat orbit. This is short enough for lower frequency microwave beamed energy. However, since we would want the beam to continue powering the vehicle out past the PowerSats orbit for some considerable distance we would want to move to a higher frequency. This higher frequency will determine the maximum orbital altitude at which the vehicle can continue to receive the beamed microwave energy.

A 91 GHz, 100 kW peak power sheet-beam klystron (74 kV, 3.6 A beam)

Moving to higher frequencies could substantially increase beam distance. A 91 GHz space rated klystron could extend the range of energy transmission out to about 20,000km. A rectenna that could receive energy at this frequency would also be needed. This is an important area for future research as it could extend our reach further out into the solar system.

Enough Energy For The Planets?

Microwave power transmission (MPT) offers a means of extracting electricity from the sun that may be beneficial not only on Earth but perhaps on the Moon and Mars as well. Recent research on behalf of the United States Department of Energy (DoE) and NASA's Jet Propulsion Laboratory (JPL) have explored the feasibility of using rectenna arrays as part of a remote lunar power system.

A study spearheaded by Professor Zoya Popvic of The University of Colorado at Boulder, for example, investigated the feasibility of creating a multi-kilowatt wireless power system to transfer power between lunar base facilities. In the conceptual system, four transmission towers power a total of five load stations, at power levels of 10 kW and distances ranging from 0.5 to 2.0 km. The feasibility study shows a better than 30-percent cost savings by using MPT technology rather than conventional transmission cables.

Chapter Eleven

The Business Case

Increasing energy consumption, shrinking resources and rising energy costs will have significant impact on our standard of living for future generations. In this situation, the development of alternative, cost effective sources of energy has to be a priority. It is possible to design, engineer and deploy the proposed PowerSats within just a few years. The basic technology already exists and this technology when combined with the innovation offered by the author makes SSP an economically attractive technology for near future energy production. In the future new heavy lift launch vehicles, such as Space Launch Vehicle (SLS) could deploy ever larger versions of these PowerSats, possibly of gigawatt scale.

THE MONEY:

"Worldwide more than a trillion dollars a year goes to the energy industry, and utilities routinely construct multibillion-dollar power plants. The energy industry has a bigger wallet than the entire U.S. federal discretionary budget. Money is not directly the problem here; profitability is. The two essential factors in the cost equation are the cost per delivered watt of the solar power components, and the cost per delivered watt of getting those components to their final destination in space." [27]

Collecting solar energy in space and transmitting it to earth offers a significant untapped energy resource. The sun's energy is almost continuously available to a satellite located in a GEO orbit about the earth or in Sun-synchronous orbits. NASA has adopted an allowable cost of 5 cents/kW-hr as its target goal for competitive terrestrial power production. This choice sets the revenue stream level for a 1.2-GW facility. Once the revenue stream is known, the net present value of this revenue stream can be computed.

A simplified calculation was made for the required return on investment, assuming zero operating costs and a 40-year operating period. The calculation demonstrates the importance of strengthening the cost analysis for the operational system. For instance, using a 10 percent rate of return, $5 billion is available for the entire system. However, a 'peak power supply" would command a higher price of 10 cents/kW-hr thereby increade the revenue stream to $20 billion. For every one gigawatt rating SPS will generate 8.75 terawatt-hours of electricity per year, or 175 TW•h over a twenty-year lifetime. With current market prices of $0.22 per kW•h (UK, January 2006) and an SPS's ability to send its energy to places of greatest demand this would equate to $1.93 billion per year or $38.6 billion over its lifetime.

---

[27] Arthur P. Smith, Ph.D., The Case for Solar Power From Space From Ad Astra, Volume 16 Number 1, 2004

The example 4 GW 'economy' SPS could therefore generate in excess of $154 billion over its lifetime. The selling price of electrical power varies with time. The economic viability of space solar power is maximum if the power can be sold at peak power rates, instead of baseline rates. A significant barrier is the initial investment required before the first power is returned.

Peak Power:

Current literature discusses Space Solar Energy in terms of base load power, however, Space Solar Energy could and mostly likely will, at least initially, be marketed as Peak Power. Peak power refers to the time of day when there is the most demand for electricity, requiring more power from the electrical grid. Some plans for creating a more energy-efficient infrastructure call for power plants which are only online during peak times. Peak power can sell for twice the price of base load power. Where wholesale base load power might sell for 5 to 8 cents per Kwh. Peak power can command prices of 10 to 16 cent per Kwh, or twice the cost of NASA's estimate of 5 cent per Kwh for space solar power in its past studies and therefore generate twice the annual revenues.

It is unlikely that PowerSats will be deployed at base-load power rates. This is especially true of some constellation designs which might experience high shadowing, such as, Molniya or LEO constellations. For LEO constellations with shadow times as high as 40%, the ability to sale the power at twice the price comes in handy. The cost tracks demand: when demand is low, at night, only the low-cost baseline production is required, while when demand is high, higher-cost peaking-power supplies are brought on line to fill the demand. Inadequate spinning reserve requires load shedding by the utility, with consequent loss of revenue, or else results in temporary "brown-out" conditions and loss of frequency regulation.

To avoid this, electricity can be purchased on the spot market. Instantaneous spot-market electricity prices can skyrocket to very high values, an order of magnitude higher than baseload prices, due to instantaneous demand, but in general these price spikes are short lasting, and not easily predictable.

World demand for energy is growing rapidly. Since greater supplies from our present sources are made available only at increasing cost and, most often, with even greater environmental impact, we must seek new sources that will be sustainable, economical, and which will not exacerbate environmental damage.

Financing:

SSP is of course an energy project - not a "space project". And fundamentally there are lots of funds for energy research. For example, in Japan nuclear energy research receives several $billion per year. The "Monju" fast-breeder reactor suffered an extremely expensive accident has already cost many tens of $billions - and the budgeted cost of the plutonium re-processing plant to supply it has recently doubled to $15 billion, although the capacity has been cut in half. In addition, the electricity companies of Japan alone invest several $10s of billions per year in electricity generation and transmission equipment.

It is envisioned that each nation state would lease part of a constellation of such satellites, thereby making financing straight forward. The missing link would be the "construction financing", however, it would be possible to "turn-over" the construction financing using a "take-out mortgage". In other words, as each satellite is build the construction financing would be paid off using the permanent "mortgage financing" thereby allows the construction financing to be recycled. This works better with smaller PowerSats than with larger ones.

One of several possible financing solutions is to lease the satellites to nations or groups of nations. Leasing is considered "off balance sheet" finance, i.e., it would not normally be considered "public finance". This is because the revenues produced by the leased satellite would be sufficient to pay the lease cost. Therefore, it would be considered "revenue financing". Lease financing allows the satellite constellation developer to "pass through" the credit of each nation or group of nations. This can provide both access to capital and low cost of capital.

Some things can be financed easier than others and space is no exception. For example, it is much easier to finance the construction of a home with a mortgage in place than to finance one based on speculation, i.e., no mortgage in place. Leases are good ways to finance things like cars and trucks and buildings. Items that are reusable and have value over time are fairly easy to lease finance.

Attempting to persuade investors to risk enough capital to finance the construction of a very large space development project would run up against the same capitalization problems now faced by entrepreneurs seeking capital for ordinary space development projects such as launching communication satellites. Investors and lenders seek to maximize economic returns from capital while avoiding risk. The cost of capital is higher for riskier investments. Persuading investors and lenders to part with their capital requires making credible promises that they will receive better returns than they would have received from making alternative investments during the same time period commensurate with risk.

While investors often accept higher levels of risk than do lenders, they do so in the expectation of even better returns. Ordinary space development projects confront not only the risks that their businesses might not make money and that the technology might fail to work as projected, but also that they might not attract enough investment because the necessary capital investment is too "chunky."

In other words, the "up-front" capital investment necessary to proceed with even an ordinary space development project tends to be relatively large and to take a relatively long time period before generating cash flows or profits.Solving this fundamental problem involves using one or more forms of direct or indirect government intervention in the capital market. The crucial difference between governments and private firms is not that governments are better at managing very large projects, but that they are better at financing very large projects. Sovereign national governments may print currency, sell or mortgage public assets, or levy taxes on property and persons within their territories. Governments may borrow from private lenders or other governments against future tax revenues or guarantee payment of loans made between private lenders and private borrowers against future tax revenues.

Governments may issue bonds backed by nothing more than their promise to redeem at face value. Governments are not liquidated when they are bankrupt. Governments may offer a wide range of direct and indirect subsidies as incentives for private investment. In effect, governments exercise the kind of power over the movements of money that is tailor made for expensive development projects. Given the problems inherent in trying to finance very large space projects with entirely private borrowing or investment, it makes sense to look to government for direct and indirect assistance.(7)

"At $100 per barrel, America's annual cost of imported oil is about $440B per year. During 6-8 years America will send to foreign countries $2.6-3.5 trillion for imported oil. Every year that we fail to act to substantially decrease our need for imported energy is $440 billion spent unnecessarily in the future. It is a starting point for the American industrialization of space and our transformation into a true spacefaring nation. "[28]

---

[28] Spacefaring America Weblog by Mike Snead (http://mikesnead.net)

While we have covered a lot of ground discussing space-base solar power (SBSP) technology, in this post I am going to talk about space asset financing from a purely commercial perspective. Many have argued in the past that the government should fund SBSP because of its high cost, however using today's technology we can substantially reduce the size and mass of an SBSP satellite considerably verse the older designs. This leads to the possibility that SBSP satellites could be privately financed rather than financed by the Government. In a World of decreasing carbon fuel assets such as coal and oil and the resulting increases in prices for those assets alternative energy solutions are needed. SBSP has the potential to fill the void left by decreasing supplies of carbon based assets.

Let's assume that we would want to build hundreds of them to off-set reduced carbon fuel supplies or simply to stop using carbon fuels since those fuels generate lots of pollution and CO_2 emissions that contribute to Global Warming. Even if you don't believe in Global Warming certainly you believe in the negative effects of air pollution, and everyone wants cheap energy. So, no need to spend any time on that subject. For SBSP to provide the Earth with abundant, clean and cheap power we need to think in terms of constellations of power satellites. In some PowerSat constellation designs a satellite might supply a single entity while in other designs a satellite might serve multiple entities. So, we create a company that will build, own and operate the PowerSats using whatever technology and operating in whatever orbit.

However, in this example we are not going to sell the power generated by the PowerSats we are simply going to lease them. The reason for this is that the lease contracts can provide a source of revenues.

Additionally, it can provide something equally important – credit. By leasing the PowerSats to a nation or group of nations we can create the credit needed to fund the development and deployment of the PowerSats.

This credit, since it can be rated based on the credit of the nations, cities, states, etc., can provide a way to capital accumulation. Leasing the PowerSats can also allow low cost financing because the interest rate could be low depending on the credit of each entity. This could have a large positive impact on SBSP economics and will result in lower cost energy. The company building PowerSats would need both research and development (R&D) funds and PowerSat deployment funds. We can think of this as being similar to building a house. You would need a construction loan and a mortgage loan. In this case the mortgage loan would be revenue bonds secured by the lease agreements.

The R&D could be structured as a loan or equity investment. In this example, it would be equity supplied via a Franchise License to each participating nation. If you set the Franchise Fee at $1 billion and assuming that R&D would cost $10 billion, we would need ten Franchisees. Each Franchisee would obtain the exclusive right to receive the energy generated and transmitted. They are not buying the energy because they would own it via the lease of the PowerSat. Lease financing could be a good way to commercially fund SBSP deployment and franchise financing could fund much of the R&D. Of course you would need a really good business plan and a very good PowerSat design before anyone is going to lease a PowerSat or buy a franchise.

Ground Assets:

While the discussion has been on space assets such as PowerSats, Spaceships, etc., ground assets also play an important role as well. Items such as launch facilities, PowerSat Factories and the rectenna sites are all needed infrastructure. Such infrastructure can be financed in many ways including private, government and public/private partnerships such as revenue bonds, tax credits, and tax increment financing.

While the proponents of space solar power have long called for a government program to implement this form of energy, history has demonstrated that that approach would be costly, waste full, unfocused and mostly likely doomed to failure due to mismanagement and political corruption. Private space efforts, while few in number, have demonstrated cost reductions of orders of magnitude less than similar government efforts.

Potential benefits of commercialization include:

- Lower development cost (removal of government overhead)
- Innovation and Value Engineering (verse cost plus with no incentive to save)
- Fixed contract (no cost overruns)
- More bidders (more competition = lower cost)
- Long-term financing (10 – 30 years)
- Lower cost financing (financing less money)
- Lower operations cost
- Expanded exploration opportunities (money saved can be invested in other programs)
- More rapid development (more programs faster and at less cost)
- Unlimited Access to Capital (Wall St.)

NASA's past analysis of the cost of commercialization has always been deliberately flawed in order to discourage such commercialization. This is anti-competitive behavior pure and simple and is an extension of corrupt political influence coming from Washington. As examples, NASA either does not include all potential cost savings such as the removal of government (NASA) overhead costs or indicates grossly high return on investment (ROI) requirements for private investment to discourage congressional support for commercialization.

As an example NASA often assumes venture capital investment with returns on investment as high as 40% as it did in the COTS program (and a paper written by Dr. Griffin years ago) – completely ignoring all other commercial financing options. The result is that the COTS program is a half-heart and hugely under funded program, by both the private sector and government with little potential for success. Surprised! Don't be, that was the intent to start with, throw the alt-space guys a bone to shut them up while moving forward with Ares 1/5 development.

Examples of exceptionally high ROI requirements can be found in every NASA study ever written on space commercialization. NASA either does not understand the potential for commercial financing or is deliberately discouraging commercial investment by making it appear to be far more expensive than it really is. My guess it that since it is often identified as a potential commercial financing option, NASA deliberately discourages its use by over estimating its cost and under estimating the potential saving. Others in government do this as well, as an example the CBO indicates investment spreads of 2.7% over government financing (treasury bonds) when in fact the spread could be as little as .5 to 1%. Also, studies often fail to include cost savings by removing government overhead cost (around 40% for NASA) and historically high cost overruns in government projects (average 70% for NASA programs) or the fact that lower cost development means you need to finance less capital to start with.

NASA also assumes short terms of 5-10 years verse 15-30 year mortgages which forces fast repayment and therefore higher annual costs for the financing payback. Part of the problem is that NASA's planning is so short term, or in the case of today – so non-existent that it is difficult to consider the benefits of long-term financing. NASA's anti-competitive practices are the only reason we have not seen billions of dollars in private sector investment in space.

It is also the number one reason we don't have people living and working of the Moon and Mars today. Some things can be financed easier than others and space is no exception. For example, it is much easier to finance the construction of a home with a mortgage in place than to finance one based on speculation, i.e., no mortgage in place.

Leases are good ways to finance things like cars and trucks and buildings. Items that are reusable and have value over time are fairly easy to lease finance. Yet NASA has continued to insist on expendable architectures that benefit few, are not cost effective and can only be financed at the expense of the taxpayer. The fundamental problem in opening any contemporary frontier, whether geographic or technological, is not lack of imagination or will, but lack of capital to finance initial construction which makes the subsequent and typically more profitable economic development possible.

Attempting to persuade investors to risk enough capital to finance the construction of a very large space development project would run up against the same capitalization problems now faced by entrepreneurs seeking capital for ordinary space development projects such as launching communication satellites. Investors and lenders seek to maximize economic returns from capital while avoiding risk. The cost of capital is higher for riskier investments. Persuading investors and lenders to part with their capital requires making credible promises that they will receive better returns than they would have received from making alternative investments during the same time period commensurate with risk. While investors often accept higher levels of risk than do lenders, they do so in the expectation of even better returns.

Ordinary space development projects confront not only the risks that their businesses might not make money and that the technology might fail to work as projected, But also that they might not attract enough investment because the necessary capital investment is too "chunky." In other words, the "up-front" capital investment necessary to proceed with even an ordinary space development project tends to be relatively large and to take a relatively long time period before generating cash flows or profits.

Solving this fundamental problem involves using one or more forms of direct or indirect government intervention in the capital market. The crucial difference between governments and private firms is not that governments are better at managing very large projects, but that they are better at financing very large projects. Sovereign national governments may print currency, sell or mortgage public assets, or levy taxes on property and persons within their territories. Governments may borrow from private lenders or other governments against future tax revenues or guarantee payment of loans made between private lenders and private borrowers against future tax revenues.

Governments may issue bonds backed by nothing more than their promise to redeem at face value. Governments are not liquidated when they are bankrupt. Governments may offer a wide range of direct and indirect subsidies as incentives for private investment. In effect, governments exercise the kind of power over the movements of money that is tailor made for expensive development projects. Given the problems inherent in trying to finance very large space projects with entirely private borrowing or investment, it makes sense to look to government for direct and indirect assistance.

CHAPTER Twelve
Conclusion

Accelerating the development of space based solar power is important to the future of mankind. A continuous and clean source of energy is greatly needed to sustain growth and for the protection of the Earth's environment. New and innovative approaches to in-space power production have the potential of offering cost effective supplies of power, more quickly than previously thought.

A new baseline model is needed that substantially reduces mass to orbit. Such an innovation in PowerSat design is the use of constellations of smaller, closer satellites. Such an approach could diversify delivery of power to multiple rectenna while keeping ground stations to acceptable size.

---

[29] Image retrieved from http://lowtestosteronecure.org/wp-content/uploads/2013/10/conclusion.jpg

The primary challenge for space solar power towers is economics and most of the economics has been related to space infrastructure and launch costs. By eliminating most of all of the past infrastructure concepts and using smaller, closer PowerSats we have a better economic structure. Economically, an SPS deployment project would create many new jobs and contract opportunities for industry, which may have political implications in the country or region which undertakes the project.

Over half the cost of SPS has been associated with launch costs. To reduce launch costs, the size of the system must be reduced. An incremental approach to PowerSat development is available by first considering their use as part of a low cost In-space transportation system. This would allow the initial deployment of small PowerSats in the megawatt power range. Certainly the energy from an SBSP would reduce political tension resulting from unequal distribution of energy supplies. The northern hemisphere is the primary consumer of energy and the proposed system serves that market much better than GEO concepts. Developing the industrial capacity needed to construct and maintain one or more SPS systems would significantly reduce the cost of other space endeavors. For example, a manned mars mission might only cost hundreds of millions, instead of tens of billions, if it can rely on an already existing and completely reusable capability.

We have shown that the basic technology does exist to build much lower mass PowerSats and that PowerSats can be used for both Earth Energy and In-space Transportation possibly even at the same time.

Any rational energy policy for the United States must support the steps needed to make that happen, including increased investment in reducing launch costs, reserving radio frequency spectrum for power transmission, and moving towards a billion dollars per year in a robust and diverse program of R&D on space solar power. Also, it is important the SSP receive the same consideration as other forms of energy in subsidies and tax credits.

We have covered a number of technologies that could be applied to PowerSat design and thereby lower the mass to orbit. These technologies include:

- Concentrated Solar Power
- Rainbow Arrays
- Closer Orbits
- Space Power Relays
- Carbon/Carbon Radiators
- Launch Vehicle Tank Radiators

The technology now exists to solve the Earth's energy problems with low cost power from space and at the same time solve the Global Warming problem. This technology will stimulate new investment in launch vehicles and can provide a basis for development of solar electric space transportation.

There is still much work to do but there is no longer any question that SSP can be an economically viable part of the Earth's energy equation and will be a major factor in moving humanity into space. SSP is the next step in space development and it is a step we can take now.

The SSP system described herein is a new baseline model that is technologically and economically feasible because the Solar Power Satellite segment of this system:

- can be built on Earth in production facilities much like communications satellites are currently produced
- can be launched with currently available satellite launch vehicles
- would operate from LEO
- would not require further assembly on-orbit by astronauts

- would utilize concentrating optics to massively concentrate solar energy to boost photovoltaic throughput
- Can be built, launched, operated and improved upon in a bootstrapping process with short time-to-term for each satellite system.

With the small ground receiver array area required, this SSP system is well suited to servicing high density regions. An operational SSP system of the design proposed herein can provide a financially lucrative and environmentally friendly, renewable electrical power resource to peak-power markets.

Authors Biography

D*anny Royce Jones, Sr. (born April 9th, 1959 in Abilene, Texas) is a technology entrepreneur, venture manager, IP developer and space technology investor, injecting both start-up capital and IP into new space related and Earth based ventures. He has a degree in Construction Management and Associates degrees in Computer Programming and Computer Aided Design. Mr. Jones is a former* **Marine Corps NCO** *with eight years of active military service and three years reserve service. He is a published author with innovative approaches to space development in the area of launch vehicle design and heavy lift vehicles and space based solar power (***SBSP***). Mr. Jones is a known innovator in field of commercial and difficult project finance. Mr. Jones has fifteen years' experience in commercial property construction and finance. He has participated in structuring financing on numerous commercial properties and is very experienced in structuring innovative commercial financing, asset backed financing, TIF financing, Tax credits and Government backed financings. He is an active real estate investor.*

Ali Baghchehsara was born February 25th, 1993 in Tehran, Iran. As a talented Aerospace Researcher, inventor, and member of the American Society of Mechanical Engineers (**ASME**), he has been researching and working on engines since he was 12 years old. In 2010 he participated in the National Iranian Aerospace Competition,  and took **first prize** for his work in the field of jet propulsion.

In 2013, he was elected to be an Associate Member of the **Sigma Xi** Scientific Research Society. Currently, he is an Aerospace Engineering student at Karaj Payam Noor University, and has successfully worked as a R&D Engineer for a large automotive company. During his university time, he has also founded and led an Aerospace Association in Alborz Province for two years.

Ali is creative person with innovative approaches to aerospace development in the areas of engine design and space based solar power (SBSP) and desires to one day become a **NASA** Astronaut. He is currently a member of several engineering associations, including the IET, SGAC, SAPOE, IFIA, UIA, MRO Asia, Innovate UK and ImechE.

References

1. KALAM-NATIONAL SPACE SOCIETY ENERGY TECHNOLOGY UNIVERSAL INITIATIVE, August 16, 2010, An International Preliminary Feasibility Study on Space Based Solar Power Stations, R. Gopalaswami; India.

2. URSI White Paper on Solar Power Satellite (SPS) Systems and Report of the URSI Inter-Commission (June 2007), Working Group on SPS, Retrieved from http://www.rish.kyoto-u.ac.jp/SPS/WPReportStd.pdf

3. Seth D. Potter, Harvey J. Willenberg, Mark W. Henley, and Steven R. Kent, ARCHITECTURE OPTIONS FOR SPACE SOLAR POWER, 1999, the Space Studies Institute.

4. A. Baghchehsara, Electro-Magnetic Regulator for Liquid Missile Motors, DOI: 10.1111/j.1945-5100.2012.01401_2.x, Volume 47, Issue s1 of Meteoritics & Planetary Science, pages A18-A19, Wiley Publication, Printed in USA, Aug 2012.

5. Glaser, P.; Maynard, O.E.; Mockovciak, J. Ralph, E.L.: Feasibility Study of a Satellite solar Power Station. NASA CR-2357, Feb 1974.

6. J.E. Drummond, Comparison of Low Earth Orbit and Geosynchronous Earth Orbit, Power Conversion Technology, Inc., 1980.

7. Jones, R: Alternative Orbits - A New Space Solar Power Reference Design, Online Journal of Space Communication, Issue 16, 2010 http://spacejournal.ohio.edu/issue16/jones.html

8. N. Komerath, N. Boechler, S. Wanis, Space Power Grid- Evolutionary Approach To Space Solar Power, Proceedings of the ASCE Earth&Space 2006 Conference, League City, Texas, April 2006.

9. Dickinson, Richard M., COMPARISON OF INTERCONTINENTAL WIRELESS AND WIRED POWER TRANSMISSION, Jet Propulsion Laboratory, California Institute of Technology. Pasadena.

    John C. Mankins. National Aeronautics and Space Administration, Washington. D.C. 7546-0001, Proceedings of the ASCE Earth&Space 2006 Conference, League City, Texas, April 2006 Space Power Grid- Evolutionary Approach To Space Solar Power.

10. Dave Criswell, Lunar Solar Power (LSP) System: Practical Means to Power Sustainable Prosperity, Search and Discovery Article #70070 (2009).

11. Krafft Ehricke, A prospect of Space program Strategy and Goals in the the Decade of the Eighties, 1980 Space Global Co.

12. Komerath, N.M., Boechler, N., "The Space Power Grid". IAC-C3.4.06, 57th International Astronautical Federation Congress, Valencia, Spain, October 2006.

13. Ivan Bekey, Richard Boudreault, An economically viable space power relay system, Published by Elsevier Science Ltd 1999.

14. Grumman Aerospace Corporation, SPACEBORNE RADAR STUDY, 1974.

15. Brown, C: Beamed Microwave Power Transmission and its Application to Space, IEEE Transactions on microwave and techniques, Vol. 40 No. 6, June 1992.

16. Mark M. Hopkins, The Satellite Power Station and Non-cost Uncertainty Aspects of Risk. The Rand Corporation, 1980.

17. Geoffrey A. Landis, Reinventing the Solar Power Satellite, National Aeronautics and Space Administration, Glenn Research Center, Cleveland, Ohio, 2004.

18. John C. Mankins, A Fresh Look at Space Solar Power: New Architectures, Concepts and Technologies, IAF-97-R.2.03, 38th International Astronautical Federation, Advanced Projects Office National Aeronautics and Space Administration, 1997.

19. National Research Council of the National Academy of Sciences, Laying the Foundation for Space Solar Power: An Assessment of NASA's Space Solar Power Investment Strategy. An evaluation of NASA 's Space Solar Power (SSP) Exploratory Research and Technology (SERT) program conducted in 1999-2000, 2001, 95 pages.

20. J.E. Drummond, Comparison of Low Earth Orbit and Geosynchronous Earth Orbit, Power Conversion Technology, Inc., 1980.

21. Mark M. Hopkins, The Satellite Power Station and Non-cost Uncertainty Aspects of Risk. The Rand Corporation, 1980.

22. Allan Kotin, Satellite Power System (SPS) State and Local Regulations as Applied to Satellite Power System Microwave Receiving Antenna Facilities, DOE/NASA, October 1978.

Acronyms

Space-based Solar Power (SBSP)

Space Solar Power (SSP)

Power Satellites (PowerSats)

Micro Electro Mechanical System (MEMS)

Wireless Power Transmission (WPT)

Space Power Relay (SPR)

Space Power Grid (SPG)

Gigawatt (GW)

National Aeronautics and Space Administration (NASA)

Department of Energy (DOE)

Low Earth Orbit (LEO)

Equatorial Medium Earth Orbit (EMEO)

Geostationary Orbit (GEO)

Kilograms (Kg)

Solar Electric Propulsion (SEP)

Beamed Solar Electric Propulsion (BSEP)

Photovoltaic (PV)

Earth-to-orbit (ETO)

Sun-synchronous Orbit (SS-O)

www.ingramcontent.com/pod-product-compliance
Lightning Source LLC
Chambersburg PA
CBHW051804170526
45167CB00005B/1868